A FIRST COURSE IN ABSTRACT ALGEBRA

VNR NEW MATHEMATICS LIBRARY

under the general editorship of

J. V. ARMITAGE
Professor of Mathematics
University of Nottingham

N. CURLE
Professor of Applied Mathematics
University of St Andrews

The aim of this series is to provide a reliable modern coverage of those mainstream topics that form the core of mathematical instruction in universities and comparable institutions. Each book deals concisely with a well-defined key area in pure or applied mathematics or statistics. Many of the volumes are intended not solely for students of mathematics, but also for engineering and science students whose training demands a firm grounding in mathematical methods.

A First Course in Abstract Algebra

P. J. HIGGINS

*Professor of Mathematics,
King's College, London.*

VAN NOSTRAND REINHOLD COMPANY

New York Cincinnati Toronto London Melbourne

© P. J. Higgins, 1975.

ISBN 0 442 30083 2 cloth
      0 442 30084 0 paperback

All rights reserved. No part of this work covered by the copyright hereon may be reproduced or used in any form or by any means — graphic, electronic, or mechanical, including photocopying, recording, taping, or information storage and retrieval systems — without written permission of the publishers.

Published by Van Nostrand Reinhold Company Limited, Molly Millar's Lane, Wokingham, Berkshire

Typeset in Great Britain by
Preface Limited, Salisbury, Wilts
and printed by
Redwood Burn Limited, Trowbridge and Esher

Preface

A student's first encounter with any new mathematical concept should ideally be closely followed by a study of the applications which justify its introduction. In the case of groups, rings and fields, which form the chief topic of this book, there are some difficulties in achieving the ideal. On the one hand, the really important applications, to such problems as the solubility of equations by radicals or the classification of surfaces, are too difficult for a first course. On the other hand, the applications that can be appreciated at an early stage are largely concerned with problems that can as easily be solved by other methods and are therefore unconvincing as a justification of the abstract algebra. As a result, the subject is often taught in isolation as an abstract discipline which the student must take on trust.

In this text, which is based on courses given to first-year students at King's College, London over a period of years, I have tried to combine a careful treatment of the rudiments of abstract algebra with a study of various topics in which the use of abstract algebra, though not essential, is natural and illuminating. Elementary number theory and the factorization of polynomials are the two main topics of this kind and they are developed alongside the abstract theory, reaching such results as Euler's theorem, unique factorization theorems for integers and polynomials, Eisenstein's criterion and the theory of partial fractions. I hope that this arrangement will help the reader to appreciate the usefulness as well as the beauty of the abstract ideas.

The main text is self-contained except in the last chapter, where the so-called Fundamental Theorem of Algebra is used without proof. The illustrative examples are, however, drawn from many parts of mathematics and involve some concepts not defined in the text. For instance, no attempt is made to define real and complex numbers accurately, but they are used freely in the examples because they will provide for many readers the most familiar material on which to try the new ideas. This in no way affects the rigour of the course, which is based on intuitive notions of set-theory and clearly-stated assumptions about integers.

I am indebted to many of my students who, through their efforts to understand abstract algebra, have helped to determine the present form of the course. I have also benefited greatly from discussions with my colleagues at King's College on the relative merits of various methods of presentation. I should like to express my grateful appreciation to all who have helped me to produce the book, and especially to Mrs. D. Woods for her prompt and efficient typing of the manuscript.

King's College,   Philip J. Higgins
London.
April 1974

Contents

PREFACE	v
1. WHAT IS ABSTRACT ALGEBRA?	1
2. SET THEORY	11
3. THE INTEGERS	31
4. GROUPS	40
5. FACTORIZATION IN Z	58
6. NEW GROUPS FROM OLD	69
7. LINEAR CONGRUENCES IN Z	82
8. RINGS AND FIELDS	95
9. THE RINGS Z_n AND THE FIELD Q	111
10. RINGS OF POLYNOMIALS	123
11. POLYNOMIALS OVER C, R, Q, AND Z	143
INDEX	156

CHAPTER 1

What is Abstract Algebra?

Algebra is the study of operations and the laws governing them. The meaning of the word "operation" will be made clear in Chapter 2, but some familiar examples will indicate the kind of operations we have in mind. The important thing to notice in these examples is that different operations obey different laws, giving rise to different types of algebra. Nevertheless, there are similarities which suggest the possibility of a common approach.

Example 1.1. *Standard algebra.*   This is the algebra that most people meet first. One is taught at an early age that, in algebra, letters denote numbers, and equations are used to express relations between them. Some equations are true for all possible values of the letters appearing in them (these are often called identities). Others are true only for some of the possible values of the letters, and the problem of determining all these values is known as "solving equations".

The operations of standard algebra are addition, subtraction, multiplication and division. Addition and multiplication operate on pairs of numbers x, y to produce their sum, $x + y$, and their product $x \times y$ (which we shall usually write as xy). Subtraction is closely related to addition and is denoted by the minus sign. This sign is customarily used in two different ways: (a) as a *binary* operation, acting on a pair of numbers x and y to give their difference, $x - y$, or (b) as a *unary* operation, acting on a single number x to give its negative, $-x$. These two operations are connected by the equations $x - y = x + (-y)$ and $-x = 0 - x$, so either of them can be expressed in terms of the other. We shall treat the unary operation as the basic one and use $x - y$ only as shorthand for $x + (-y)$. Similarly division (which is related to multiplication in the same way that subtraction is related to addition) is best thought of in terms of a unary operation (called inversion) rather than a binary one. The unary operation produces from a non-zero number x its inverse x^{-1}, and the quotient $x \div y$ can be written as xy^{-1}. In practice the division sign is almost never used by mathematicians, the notation xy^{-1} (or in some contexts x/y) being preferred to $x \div y$.

The laws of standard algebra involve the special numbers 0 and 1 as

1

2 WHAT IS ABSTRACT ALGEBRA?

well as the operations described above. The laws are as follows:

(A1) $(x+y)+z = x+(y+z)$ for all x, y, z;
(A2) $x+0 = 0+x = x$ for all x;
(A3) $x+(-x) = (-x)+x = 0$ for all x;
(A4) $x+y = y+x$ for all x, y;
(M1) $(xy)z = x(yz)$ for all x, y, z;
(M2) $x1 = 1x = x$ for all x;
(M3) $xx^{-1} = x^{-1}x = 1$ for all $x \neq 0$;
(M4) $xy = yx$ for all x, y;
(AM1) $x(y+z) = xy + xz$ and $(x+y)z = xz + yz$ for all x, y, z;
(AM2) $1 \neq 0$.

These laws are true, for example, if the letters stand for rational numbers, real numbers or complex numbers. They are used, perhaps subconsciously, in the manipulation of equations and formulae in elementary algebra, and they are almost sufficient for this purpose in the sense that most valid laws of elementary algebra can be deduced logically from them. We shall return to this particular set of laws in Chapter 8.

Example 1.2. *The algebra of polynomials.*   Consider the set of all polynomials in one indeterminate X with, say, real coefficients. They are expressions of the form $a_0 + a_1 X + a_2 X^2 + \ldots + a_n X^n$, where the a_i are real numbers. Two polynomials can be added and multiplied according to the usual rules and the result is, in each case, a polynomial. Every polynomial has a negative (change the signs of all the coefficients), and there are two special polynomials 0 and 1 (in which $a_i = 0$ for $i > 0$ and $a_0 = 0$ or 1). All the laws of standard algebra except (M3) are true for polynomials. However (M3) is not true because not all non-zero polynomials have polynomial inverses. The ones which do have inverses are those of degree zero (i.e. those for which $a_0 \neq 0$ and $a_i = 0$ for $i > 0$). For these invertible polynomials the law (M3) is valid; but if we want the law to hold for all non-zero polynomials then we must extend the algebra of polynomials to the algebra of rational functions (quotients of polynomials). These algebras will be studied in Chapter 10.

Example 1.3. *Matrix algebra.*   For simplicity we restrict our attention to the set of 2×2 matrices

$$\begin{pmatrix} a & b \\ c & d \end{pmatrix}$$

with real entries. Two such matrices can be added and multiplied according to the rules

$$\begin{pmatrix} a & b \\ c & d \end{pmatrix} + \begin{pmatrix} a' & b' \\ c' & d' \end{pmatrix} = \begin{pmatrix} a+a' & b+b' \\ c+c' & d+d' \end{pmatrix},$$

$$\begin{pmatrix} a & b \\ c & d \end{pmatrix} \begin{pmatrix} a' & b' \\ c' & d' \end{pmatrix} = \begin{pmatrix} aa'+bc' & ab'+bd' \\ ca'+dc' & cb'+dd' \end{pmatrix}.$$

These operations are not arbitrarily chosen but are dictated by the use of matrices to represent linear transformations. The claim that they are "natural" or, at any rate, mathematically interesting operations is supported by the fact that they obey most of the laws of standard algebra. There are special matrices

$$\mathbf{0} = \begin{pmatrix} 0 & 0 \\ 0 & 0 \end{pmatrix} \quad \text{and} \quad \mathbf{1} = \begin{pmatrix} 1 & 0 \\ 0 & 1 \end{pmatrix}$$

and with these replacing 0 and 1, all the standard laws except (M3) and (M4) are valid, the letters now standing for matrices. As with polynomials, not all non-zero matrices have inverses. In fact

$$\begin{pmatrix} a & b \\ c & d \end{pmatrix}$$

has an inverse if and only if $\Delta = ad - bc \neq 0$, and its inverse is then

$$\begin{pmatrix} d/\Delta & -b/\Delta \\ -c/\Delta & a/\Delta \end{pmatrix}.$$

These *invertible* (or *non-singular*) matrices do satisfy (M3), but in contrast with the case of polynomials it is not possible to extend the algebra of matrices so that (M3) is valid for *all* non-zero matrices. (Why not? Think about the equation

$$\begin{pmatrix} 1 & 0 \\ 0 & 0 \end{pmatrix} \begin{pmatrix} 0 & 0 \\ 0 & 1 \end{pmatrix} = \mathbf{0}.)$$

The fact that (M4) is not valid for matrices is easily checked by looking for a simple counter-example. The validity of (M1) can be proved by a direct calculation, but it is less mysterious when proved in the context of linear transformations.

In case it should be imagined that all algebra is concerned with operations like addition and multiplication satisfying some or all of the standard laws, we now give two examples of quite different types of algebra.

Example 1.4. *Vector algebra.* An n-dimensional real vector is a row of n real numbers, $\mathbf{u} = (u_1, u_2, \ldots, u_n)$. If $\mathbf{v} = (v_1, v, \ldots, v_n)$ then addition of vectors is defined by $\mathbf{u} + \mathbf{v} = (u_1 + v_1, u_2 + v_2, \ldots, u_n + v_n)$. This addition, with the zero vector $\mathbf{0} = (0, 0, \ldots, 0)$, satisfies the laws (A1)–(A4). Instead of multiplication of two vectors, the natural operation to consider is multiplication of vectors by scalars (i.e. real numbers). If λ is a real number, then $\lambda\mathbf{u}$ is defined to be the vector $(\lambda u_1, \lambda u_2, \ldots, \lambda u_n)$ and the other standard laws are replaced by the laws

$$\lambda(\mathbf{u} + \mathbf{v}) = \lambda\mathbf{u} + \lambda\mathbf{v},$$
$$(\lambda_1 + \lambda_2)\mathbf{u} = \lambda_1\mathbf{u} + \lambda_2\mathbf{u},$$
$$(\lambda_1\lambda_2)\mathbf{u} = \lambda_1(\lambda_2\mathbf{u}),$$
$$1\mathbf{u} = \mathbf{u},$$

which hold whenever $\mathbf{u}, \mathbf{v}$ are n-dimensional real vectors and $\lambda, \lambda_1, \lambda_2$ are real numbers. In the last equation, 1 means the real number 1. There is also a useful operation known as scalar product which operates on pairs of vectors $\mathbf{u}, \mathbf{v}$ to produce a *scalar* $\mathbf{u} \cdot \mathbf{v}$. It is defined by $\mathbf{u} \cdot \mathbf{v} = u_1 v_1 + u_2 v_2 + \ldots + u_n v_n$ and is used to describe angles between vectors.

If we look for products of the type encountered in the first three examples, that is, operations on pairs of vectors such that the product is again a vector, we find several interesting examples. In two dimensions we can define $\mathbf{uv} = (u_1 v_1 - u_2 v_2, u_1 v_2 + u_2 v_1)$ (imitating the formal multiplication of $(u_1 + iu_2)$ by $(v_1 + iv_2)$ with $i^2 = -1$) and we find that *all* the laws of standard algebra are valid, the inverse of $\mathbf{u}$ being $(u_1/(u_1^2 + u_2^2), -u_2/(u_1^2 + u_2^2))$ if $\mathbf{u} \neq (0, 0)$. The result is the algebra of complex numbers.

In three dimensions there is the well-known *vector product* defined by $\mathbf{u} \times \mathbf{v} = (u_2 v_3 - u_3 v_2, u_3 v_1 - u_1 v_3, u_1 v_2 - u_2 v_1)$. Far from satisfying the laws (M1)–(M4) this product satisfies $\mathbf{u} \times \mathbf{v} = -\mathbf{v} \times \mathbf{u}$ and $(\mathbf{u} \times \mathbf{v}) \times \mathbf{w} + (\mathbf{v} \times \mathbf{w}) \times \mathbf{u} + (\mathbf{w} \times \mathbf{u}) \times \mathbf{v} = \mathbf{0}$ for all 3-dimensional vectors $\mathbf{u}, \mathbf{v}, \mathbf{w}$; there is no "1" and there are no inverses. Other interesting examples occur in four dimensions (Hamilton's quaternions which satisfy (M1)(M2) and (M3)) and in eight dimensions (Cayley's octonions which satisfy (M2) and (M3)).

Example 1.5. *The algebra of sets.* Let S be a fixed set and consider all its subsets, including the empty set $\emptyset$ and S itself. If A and B are subsets of S then $A \cup B$ and $A \cap B$ are also subsets; so is A', the complement of A in S. (If these notions are not familiar, see the beginning of Chapter 2 for their definitions.) The three operations $\cup, \cap, '$, and the special subsets $\emptyset$ and S satisfy the following laws, valid for all subsets A, B and C of S.

$$A \cup B = B \cup A, \quad A \cap B = B \cap A,$$
$$(A \cup B) \cup C = A \cup (B \cup C), \quad (A \cap B) \cap C = A \cap (B \cap C),$$
$$A \cap (B \cup C) = (A \cap B) \cup (A \cap C), \quad A \cup (B \cap C) = (A \cup B) \cap (A \cup C),$$
$$A \cap (A \cup B) = A, \quad A \cup (A \cap B) = A,$$
$$A \cup \emptyset = A, \quad A \cap S = A,$$
$$A \cup S = S, \quad A \cap \emptyset = \emptyset,$$
$$A \cup A' = S, \quad A \cap A' = \emptyset,$$
$$(A')' = A,$$
$$(A \cup B)' = A' \cap B', \quad (A \cap B)' = A' \cup B'.$$

These are the laws of *Boolean algebra*, named after George Boole who introduced algebraic methods into the study of logic and the treatment of syllogisms. Boole's notation was different, but equivalent, to the above, and he was concerned not with sets but with the truth of propositions. The connection between sets and logic arises as follows. Let $P(x)$, $Q(x)$, etc. denote propositions which contain a variable x whose values range over the members of a set S. Thus each proposition is either true or false when a particular member of S is substituted for x. If we denote by $\Gamma(P)$ the set of all $x \in S$ for which $P(x)$ is true, then we have $\Gamma(P \text{ or } Q) = \Gamma(P) \cup \Gamma(Q)$, $\Gamma(P \text{ and } Q) = \Gamma(P) \cap \Gamma(Q)$, and $\Gamma(\text{not } P) = \Gamma(P)'$. Thus the logical connectives "or", "and" and "not" are closely related to the operations $\cup$, $\cap$ and $'$ on sets. In fact the laws of Boolean algebra set out above are essentially the same as the laws of logic, and there is a good case for maintaining that Boolean algebra is more fundamental than standard algebra. Boolean algebra has many applications including the design of switching circuits for computers. Readers interested in an elementary account of this application should read the last chapter of *Topics in Algebra* by Hazel Perfect (Pergamon). Boole's original treatise *The Mathematical Analysis of Logic* (Cambridge 1847, reprinted by Blackwell, Oxford, 1948) still makes fascinating reading and is not difficult to understand.

From all these examples it appears that there is a wide variety of operations susceptible to algebraic treatment and that they obey laws which also vary considerably. However certain types of law seem to crop up again and again, for example the *commutative laws* $x + y = y + x$, $xy = yx$, $A \cup B = B \cup A$, $A \cap B = B \cap A$, the *associative laws* $(x + y) + z = x + (y + z)$, $(A \cup B) \cup C = A \cup (B \cup C)$, etc. and the *distributive laws* $x(y + z) = xy + xz$, $A \cup (B \cap C) = (A \cup B) \cap (A \cup C)$ and $A \cap (B \cup C) = (A \cap B) \cup (A \cap C)$. Note that these laws are not automatically true for all operations. For example the associative law of subtraction, $(a - b) - c = a - (b - c)$, is not true in standard algebra, nor is the distributive law of addition over multiplication, $a + (bc) = (a + b)(a + c)$.

We come now to one of the most far-reaching ideas of modern mathematics — *abstraction*. There are two alternative ways of thinking about operations and their laws: (a) we can think of the laws as true statements about particular operations on particular objects, or (b) we can think of the laws as the rules of a game to be followed blindly without reference to the nature of the objects being operated on, the aim being to deduce new laws by purely logical methods from the given ones. For example, we are following method (a) if we study specifically the algebra of real numbers or the algebra of 2 × 2 real matrices. On the other hand, if we study only the logical consequences of the laws of standard algebra, we are following method (b). That the two methods are in fact different can easily be illustrated by the sentence "the equation $x^2 = 2$ has a solution". This statement is true in the algebra of real numbers, but false in the algebra of rational numbers. Since the laws of standard algebra are true in the algebra of rational numbers it follows that the existence of a solution of $x^2 = 2$ is not a logical consequence of these laws; thus there are true statements about real numbers which are not theorems of standard algebra.

Abstract algebra is the study of operations and laws by method (b). It is a good example of the axiomatic method, which is the chief characteristic of twentieth century mathematics. In the axiomatic method one assumes certain statements (variously known as postulates, hypotheses or axioms) about unspecified objects, and examines their logical consequences. In our case the unspecified objects are the operations and the entities on which they act, and the axioms are the laws we wish to study. The method has several advantages over the study of specific operations, advantages which it is the aim of this book to exploit in an elementary context.

The first advantage of abstract algebra is *generality*: any statement that can be deduced from a given set of laws will be true generally in any algebraic context where these laws are valid. This leads to great economy of proof — one proof will serve to establish similar theorems in many contexts. The second advantage, *flexibility*, reinforces this generality. One is free to select particular sets of laws for special consideration and to reject those laws that seem irrelevant. For example, any consequence of the standard laws that can be deduced *without using (M3) or (M4)* will be true not only for real numbers, rational numbers and complex numbers, but also for integers, for polynomials and for matrices. This suggests that the set of standard laws with (M3) and (M4) deleted is worth studying in its own right. One is also free to study operations one at a time, even though in particular contexts they tend to occur in bunches. The third and most important advantage is *clarity*. The phenomenon of "not seeing the wood for the trees" is particularly common in mathematics, even among the experts. The history of mathematics provides many an instance of a difficult and obscure theorem in a special branch of the subject which is later seen to

AXIOMATIC METHOD   7

be a special case of a general principle of astonishing simplicity and wide application. The point is that one grows accustomed to an accumulation of facts about particular mathematical objects and, when faced with a problem, one tends to use the first fact that comes to mind and seems to be relevant. It is often only when one tries to do without some of these facts (either through necessity in a new context or as an act of discipline) that one sees a simpler argument based on more fundamental principles. The axiomatic method, by throwing away hypotheses whenever possible and basing its arguments on the minimum of information, helps to lay bare the logical connections between mathematical facts to an extent not otherwise possible.

For those readers who need convincing that abstract algebra is worth pursuing or who think that it may prove impossibly difficult, it should be pointed out that they have probably used the method implicitly themselves for some years. When manipulating formulae and equations in elementary algebra, do they at every step remember that the symbols stand for (say) real numbers and use this fact in thinking what to do next? Or do they perform the manipulation mechanically without regard to the meaning of the letters, using only certain rules which have become second nature to them? The fact that some of them use the *wrong* rules only confirms that they are really doing abstract algebra.

To end this introductory chapter we shall give some simple examples of logical deductions from algebraic laws. They are intended to give a hint of the flavour of the subject and will be written out in full detail, stating at each stage which law or laws have been used. As the book progresses this sort of detailed proof will be gradually replaced by a less formal and more readable style in which the main steps are given but some details are left to the imagination. The reader is strongly recommended to put in all the details himself until such time as he is completely confident of his ability to argue correctly and knows which details can safely be omitted.

We shall take as axioms the laws of "standard algebra" as formulated in Example 1.1 and deduce from them the following statements:

(i)   if $a + b = a$ then $b = 0$;

(ii)   if $a + b = 0$ then $b = -a$ and $a = -b$;

(iii) $-(-a) = a$ for all a;

(iv) $a \cdot 0 = 0$ for all a;

(v) $a \cdot (-b) = -(ab)$ for all a, b;

(vi) $(-1) \cdot (-1) = 1$.

Of course, once we have deduced one of these statements from the laws we may use it in deducing the other statements, but we must avoid circular arguments such as using (i) to deduce (ii), (ii) to deduce (iii) and (iii) to deduce (i).

Proof of (i). Suppose that $a + b = a$. Then $(-a) + (a + b) = (-a) + a = 0$ by (A3). But

$$(-a) + (a + b) = ((-a) + a) + b \quad \text{by (A1),}$$
$$= 0 + b \quad \text{by (A3),}$$
$$= b \quad \text{by (A2).}$$

It follows that $b = 0$.

Proof of (ii). Suppose that $a + b = 0$. Then

$$-a = (-a) + 0 \quad \text{by (A2),}$$
$$= (-a) + (a + b) \quad \text{by hypothesis,}$$
$$= ((-a) + a) + b \quad \text{by (A1),}$$
$$= 0 + b \quad \text{by (A3),}$$
$$= b \quad \text{by (A2).}$$

A similar argument (write it down) proves that $-b = a$.

Proof of (iii). Let $b = -a$. Then $a + b = 0$, by (A3). It follows, by (ii), that $a = -b = -(-a)$.

Proof of (iv). We use the distributive law (AM1) to connect additive and multiplicative properties of 0. Let $b = a \cdot 0$. Then

$$b + b = a \cdot 0 + a \cdot 0$$
$$= a \cdot (0 + 0) \quad \text{by (AM1),}$$
$$= a \cdot 0 \quad \text{by (A2),}$$
$$= b.$$

It follows, by (i), that $b = 0$.

Proof of (v). This is very similar. We have

$$a \cdot (-b) + a \cdot b = a \cdot ((-b) + b) \quad \text{by (AM1),}$$
$$= a \cdot 0 \quad \text{by (A3),}$$
$$= 0 \quad \text{by (iv).}$$

Hence $a \cdot (-b) = -(a \cdot b)$ by (ii).

Proof of (vi). Putting $a = -1$ and $b = 1$ in (v), we have

$$(-1) \cdot (-1) = -((-1) \cdot 1)$$
$$= -(-1) \quad \text{by (M2),}$$
$$= 1 \quad \text{by (iii).}$$

It should be observed that we have not used the laws (A4), (M3), (M4) or (AM2) in these proofs.

Exercises

1. From the laws of standard algebra deduce the following statements:

   (i) $(ab)c = (cb)a$, for all a, b, c;

   (ii) if $a + c = b + c$ then $a = b$;

   (iii) $a^2 - b^2 = (a - b)(a + b)$ (where a^2 means $a \cdot a$);

   (iv) if $ab = 0$ then $a = 0$ or $b = 0$.

2. Given two binary operations $\vee$ and $\wedge$ satisfying the laws $a \vee (a \wedge b) = a$ and $a \wedge (a \vee b) = a$ for all a and b, prove that $x \vee x = x \wedge x = x$ for all x.

3. Prove that the operation $*$ defined on *real numbers* by the rule
   $$a * b = a + b + ab$$
   is commutative ($a * b = b * a$) and associative (($a * b) * c = a * (b * c)$). Is there a real number e such that $e * a = a * e = a$ for all a? If $a * b = a * c$ is it necessarily true that $b = c$?

4. Deduce from the laws of Boolean algebra that
   $$(A \cap B') \cup (A \cap (B \cup C)) = A \text{ for all } A, B, C.$$
   (You must not assume that A, B, C are sets. They can be any objects admitting operations $\cup, \cap, '$, which satisfy the laws displayed in Example 1.5. But it may help in constructing a proof to think of sets at first.)

5. Deduce from the laws of Boolean algebra that if $A \cup B = A$ then $A \cap B = B$.

6. Suppose that objects $A, B, C, \ldots$ satisfy the laws of Boolean algebra with respect to operations $\cup, \cap, '$. Write $A + B = (A \cup B) \cap (A' \cup B')$ and write 0 for $\emptyset$. Prove that the laws (A2), (A3), (A4) of standard algebra are valid, and identify $-A$. Interpret $A + B$ (i) for sets and (ii) for propositions (where $\cup$ means "or", $\cap$ means "and", and $'$ means "not").

7. (Harder). Continuing exercise 6, prove that the law (A1) is valid. Also, writing AB for $A \cap B$, and 1 for S, prove that the laws (M1), (M2), (M4) and (AM1) are valid.

8. (Harder). Which of the following statements are logical consequences of the laws of "standard algebra"?

(i) If $x^2 = y^2$ then $x = y$ or $x = -y$.
(ii) If $x^2 + y^2 = 0$ then $x = y = 0$.
(iii) If $x^2 + y^2 = a$ then $a = b^2$ for some b.

(Hint: in order to show that a statement is not a consequence of a given set of laws, you must produce an example of a system satisfying these laws in which the statement is false.)

CHAPTER 2

Set Theory

In proving facts about algebraic systems there are two mathematical tools we need apart from pure logic. In the first place we need elementary properties of integers; for example the notations x^2, x^n for powers of x presuppose knowledge of the integers $2, n$, and the law $x^m x^n = x^{m+n}$ presupposes knowledge of how to add integers. Clearly we cannot get far without such knowledge, even in abstract algebra. In the second place, for less obvious reasons, we need the elementary theory of sets. The language of set theory is used in most branches of mathematics as a medium for making precise statements and for defining basic concepts in a clear and unambiguous way. Its suitability for this purpose will become apparent later in the book. In addition, the notion of set sometimes appears in the statement of theorems; for example it can be proved that if a *finite* set of objects satisfies all the laws of standard algebra then the number of objects in the set must be a power of a prime number. This is a property of the *set* of objects, not of the individual objects, and for this reason cannot be deduced from the standard laws by logic alone.

The properties of integers needed for the development of algebra will be described in Chapter 3. In the present chapter we discuss the necessary set theory. Both these topics can themselves be treated axiomatically, but this procedure introduces complications which are inappropriate in an introductory course. Instead, we shall assume informally various well-known facts about sets and integers and, armed with this essential information, we shall proceed thereafter with a high degree of logical rigour.

For the benefit of students who have not studied set theory before, we start from the beginning and give all the definitions. This makes the chapter long, and it may be advisable to read it through first rather quickly, but carefully enough to absorb the basic concepts of *function, operation* and *equivalence relation*. If these concepts are clear the student can safely continue, referring back to Chapter 2 for further details when necessary.

A *set* is a collection of objects, considered as a single entity. Thus a football team is a set of players and a library is a set of books. The objects of which a set is composed are called its *members* or *elements* and we write $x \in S$ to mean that the object x is a member of the set S. We also say that x *belongs to S*. Two sets are to be thought of as equal (i.e. as the same set) if and only if they have the same members. Thus in speaking of a library as a set of books we ignore such features as the particular order in which the books appear on the shelves. We shall see later how to take into account such additional structural features which a set may possess, but to begin with we must consider abstract sets determined entirely by their members.

Particular sets can be specified in two ways. We can list their members, in which case we write the individual members in braces; for example $\{0, 1, \pi\}$ denotes the set with exactly three members 0, 1 and π. Alternatively, we can give a blanket description of the members by means of a characteristic property p shared by them and by nothing else. The appropriate notation in this case is $\{x; p(x)\}$ which is to be read "the set of things x for which the statement $p(x)$ is true".

Example 2.1. Certain sets occur so often in mathematics that special symbols are reserved for them. We shall, for example, always use **N** to denote the set of all natural numbers. Its members are the numbers $0, 1, 2, \ldots$ used for counting. The set of *all* integers $0, \pm 1, \pm 2, \ldots$ will be denoted by **Z** (after the German "Zahl"). Other sets which will feature frequently in examples are:

Q, the set of rational numbers (quotients of integers);
R, the set of all real numbers;
C, the set of all complex numbers.

Given a set S, we can form *subsets* of S by selecting some (or all, or none) of the members of S and making them the members of a new set. Thus a set T is called a subset of S if every member of T is a member of S. In particular S itself is a subset of S. It is very useful in this connection to introduce an *empty* set with no members at all. Since two empty sets have precisely the same members, all empty sets are equal, so we can speak of *the* empty set and use a standard symbol $\emptyset$ for it. This curious set is then a subset of every set. We use the notation $T \subset S$ or $S \supset T$ to denote that T is a subset of S (and read "T is contained in S" or "S contains T"). Note that if $S \subset T$ and $T \subset S$ then $S = T$. Usually, if we want to specify a particular subset of S we do so by stating some property distinguishing its members from the other members of S. Thus in the set of all men we have the subset of one-legged men. The notation for subsets described in this way is $\{x; x \in S \text{ and } p(x)\}$, where $p(x)$ is some statement about x. We also write, more economically, $\{x \in S; p(x)\}$ and read this as the "set of all x in S such that $p(x)$ is true". Of course, if $p(x)$ is false for *every* $x \in S$, then $\{x \in S; p(x)\} = \emptyset$.

Example 2.2. All the sets described in Example 2.1 are subsets of the set $\mathbf{C}$ of all complex numbers. Indeed they form a chain of subsets, each a subset of the next: $\mathbf{N} \subset \mathbf{Z} \subset \mathbf{Q} \subset \mathbf{R} \subset \mathbf{C}$.

Example 2.3. The set $\{x \in \mathbf{R}; 2x^2 - 4x + 1 = 0\}$ has exactly two members, $1 + \frac{1}{2}\sqrt{2}$ and $1 - \frac{1}{2}\sqrt{2}$. The sets $\{x \in \mathbf{Q}; x^2 = 3\}$ and $\{x \in \mathbf{R}; x^2 - x + 1 = 0\}$ are both empty.

If A and B are two subsets of S, their *union* $A \cup B$ is the subset $\{x; x \in A \text{ or } x \in B\}$. It is important to note that the word "or" is used here, as always in mathematics, in its inclusive sense; that is, $A \cup B$ consists of all elements of S which either belong to A or belong to B *or belong to both*. Thus both A and B are subsets of $A \cup B$ and in fact $A \cup B$ is the smallest subset of S which contains both A and B as subsets, in the sense that if C is a subset of S and $C \supset A$ and $C \supset B$, then $C \supset A \cup B$. Similarly the *intersection* of A and B is the subset $A \cap B = \{x; x \in A \text{ and } x \in B\}$. It is a subset both of A and of B and is the largest set which is a subset of both. Of course, if A and B have no common members, then $A \cap B = \emptyset$, and this alone is sufficient justification for introducing the empty set since otherwise the intersection would not always be defined and the laws of set theory would be correspondingly more complicated. The union, and intersection of several subsets $A_1, A_2, \ldots, A_n$ of S are similarly defined and we use notations

$$\bigcup_{i=1}^{n} A_i = A_1 \cup A_2 \cup \ldots \cup A_n$$

and

$$\bigcap_{i=1}^{n} A_i = A_1 \cap A_2 \cap \ldots \cap A_n.$$

For arbitrary collections of subsets (possibly infinitely many of them) we may use suffixes which are not integers but come from some set I (called an indexing set). We then write

$$\bigcup_{i \in I} A_i$$

for the union of all the sets A_i and

$$\bigcap_{i \in I} A_i$$

for their intersection. If $A \subset S$, the *complement* of A in S is defined to be $x \in S; x \notin A$, where $\notin$ means "does not belong to". The complement is denoted by $S \setminus A$ or, if S is given and fixed, by A'. The laws satisfied by union, intersection and complement have already been stated in Example 1.5.

The reader should satisfy himself that he understands the logical basis for such laws as $A \cup \emptyset = A$, $A \cap \emptyset = \emptyset$ and $\emptyset' = S$.

We have seen that the intersection, union and complement of subsets of a set are closely related to the logical connectives "and", "or" and "not". We now introduce some more logical notation and explore its relationship with set theory. The *quantifiers* $\forall$ and $\exists$ are symbols meaning roughly "for all" and "there exists", used in the following way. If $p(x)$ is a statement (or proposition) concerning a variable object x, then $(\forall x)\, p(x)$ means "$p(x)$ is true for all x" and $(\exists x)\, p(x)$ means "there exists at least one object x for which $p(x)$ is true". Without qualification the phrase "for all x" has no clear meaning; we need a description of the sort of x to be considered. This is provided by specifying a set S, the *range* of x, and restricting x to stand for an arbitrary member of S. Sometimes S is fixed once and for all (and called the universe of discourse). In this case the meaning of $(\forall x)\, p(x)$ and $(\exists x)\, p(x)$ is clear. If the range of x is liable to change from time to time we must mention it explicitly, so we write $(\forall x \in S)\, p(x)$ and $(\exists x \in S)\, p(x)$, meaning "$p(x)$ is true for all members x of S" and "$p(x)$ is true for at least one member x of S". Note that $(\forall x \in S)\, p(x)$ says the same thing as $\{x \in S; p(x)\} = S$. Also $(\exists x \in S)\, p(x)$ says the same thing as $\{x \in S; p(x)\} \neq \emptyset$. If we write $\neg$ for "not", so that $\neg p(x)$ means "$p(x)$ is false", then the statement

$$\neg(\exists x \in S)\, p(x)$$

means "there does not exist an x in S for which $p(x)$ is true". This is the same as "$p(x)$ is false for every x in S" which, in symbols, is

$$(\forall x \in S)\, \neg p(x).$$

Similarly the statement

$$\neg(\forall x \in S)\, p(x)$$

is equivalent to the statement

$$(\exists x \in S)\, \neg p(x).$$

Thus the negation symbol may be moved past a quantifier provided that we change $\forall$ to $\exists$ or *vice versa*. As further examples of the use of quantifiers we note the equations

$$\bigcup_{i \in I} A_i = \{x \in S; (\exists i \in I)(x \in A_i)\},$$

$$\bigcap_{i \in I} A_i = \{x \in S; (\forall i \in I)(x \in A_i)\},$$

which can be construed as defining the union and intersection of collections of subsets of S.

The symbol ⇒ means "implies" and is used grammatically as a verb. If p and q, are propositions then $p \Rightarrow q$ is also a proposition (p implies q) and it means "if p is true then so is q". Logically this is equivalent to "either p is false or q is true", so $p \Rightarrow q$ means the same as $((\neg p) \text{ or } q)$. The most frequent use of logical implication is between propositions involving a variable, and in this context its meaning is more natural and more interesting. For example, the square of every even integer is again even and this statement can be written $(\forall x \in \mathbf{Z})(x \text{ even} \Rightarrow x^2 \text{ even})$. In general, if $p(x)$ and $q(x)$ are two statements about x, the statement $(\forall x)(p(x) \Rightarrow q(x))$ means that $q(x)$ is true for every value of x which makes $p(x)$ true. This statement can also be written in set-theoretic language in the form $\{x; p(x)\} \subset \{x; q(x)\}$, showing the close connection between logical implication and the notion of subset. The symbol ⇐ means "is implied by" and the symbol ⇔ means "implies and is implied by". If two statements $p(x)$, $q(\text{x})$ are such that $(\forall x)(p(x) \Leftrightarrow q(x))$ is true, then we say that they are equivalent statements about x. The two sets $\{x; p(x)\}$ and $\{x; q(x)\}$ are equal in this case.

We come now to the first really important idea in set theory, that of a *function*. The idea is in essence an old one arising from consideration of the dependence of one physical quantity on another. From real-valued functions of a real variable (which the reader will have used in elementary calculus) the meaning of the word "function" has been extended over the years to include complex-valued functions, functions of several variables and vector-valued functions (for example the position vector of a particle as a function of time) and so on. A specific function is usually given by a formula from which its values can, in principle, be computed; but in developing the calculus it is necessary to speak of an arbitrary function without specifying which particular function is intended. For this to have meaning there must be an agreed criterion for deciding what is a function and what is not. The most important common feature of the various types of functions mentioned above is that *the value of a function is determined uniquely by the value of the variable or variables*. This property is now accepted as the appropriate one to use in defining the meaning of "function". The language of set theory makes it easy to cope with different types of variables and with the different types of values which a function may take. One simply needs to specify two sets, the first having as members all possible values of the variable and the second having as members all values of the function. The definition which finally emerges is the following:

Definition. Let X and Y be sets. A *function from X to Y* is given by a rule which determines, for each member of X, a corresponding member of Y. The set X is called the *domain* of the function and the set Y is called its *range*.

If f denotes a function from X to Y and if $x \in X$, we shall write $f(x)$ for the result of applying the function to the particular member x of X. Then $f(x)$ is a member of Y and is called the *value of f at x*. Note carefully that the requirements for a function are (i) that f can be applied to *every* member of X and (ii) that for each $x \in X$, $f(x)$ is a *uniquely* determined member of Y.

Example 2.4. We can define a function f from **R** to **R** by the rule: $f(x) = x^2$. This is the "squaring function". However if we try to define a "square-root" function g from **R** to **R** by the rule: "$g(x)$ is the real number whose square is x" then g fails to be a function on two counts. Firstly, if x is negative, there is no real number whose square is x, so the rule cannot be applied to x. Secondly, if x is positive, there are *two* real numbers whose square is x, so $g(x)$ is not uniquely determined. We can, however, obtain a square-root function by first restricting attention to values of x satisfying $x \geqslant 0$ and then giving a rule which picks out, for each such x, *one* of the square-roots of x. The simplest rule is to pick always the *positive* square-root (or 0). The positive square-root of x is denoted by $\sqrt{x}$ and we now have a function $\sqrt{}$ from X to **R**, where X is the set $\{x \in \mathbf{R}; x \geqslant 0\}$. The same rule can be used to define a function from X to X.

Example 2.5. We can define a function f from **Z** to **Z** by the rule:

$$\begin{cases} f(n) = -1 & \text{if } n \text{ is odd,} \\ f(n) = 1 & \text{if } n \text{ is even.} \end{cases}$$

The specification of a function by enumerating cases in this way is a common device. An extreme form of this method is to list the values of the function for all values of the variable. Thus, if X is the set $\{1, 2, 3\}$ we can define a function F from X to X by the rule

$$\begin{cases} F(1) = 3, \\ F(2) = 2, \\ F(3) = 2. \end{cases}$$

Two rules defining functions from X to Y may both have the same effect on each member of X. In this case we wish to consider them as defining the same function. If f, g are functions from X to Y we therefore say that f and g are the same function if and only if $f(x) = g(x)$ for every $x \in X$. We write $f = g$ in this case.

Example 2.6. The functions g, h from **Z** to **Z** defined by the rules: $g(n) = (-1)^n$; $h(n) = \cos n\pi$, are equal to each other and to the function f defined in Example 2.5 above.

Example 2.7. If $X = \{1, 2, \ldots, m\}$ and $Y = \{1, 2, \ldots, n\}$ then the number of distinct functions from X to Y is n^m since each function f is determined by the m values $f(1), f(2), \ldots, f(m)$, and each of these can be chosen arbitrarily from the n members of Y.

A function f from X to Y can be pictured as "sending" each member x of X to the corresponding member $f(x)$ of Y, thereby "mapping" X into Y.

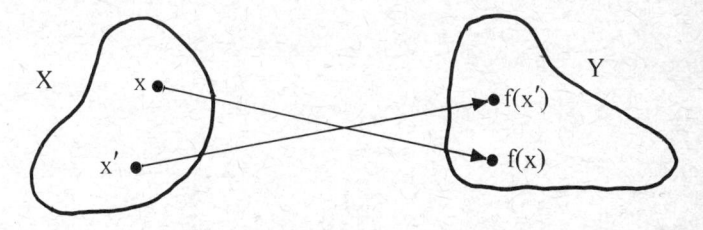

For this reason the word "mapping" or "map" is often used instead of "function" with precisely the same meaning, and the notation $f : X \to Y$ is used to signify that f is a mapping from X to Y. A different arrow $\mapsto$ is used to show the effect of the mapping or function on the individual elements of X, so that $f : x \mapsto y$ means "f sends x to y", that is, $f(x) = y$. This arrow is convenient for specifying functions; for example, the function $f : \mathbf{R} \to \mathbf{R}$ defined by $f(x) = x^3 + x$ can be described as the function $x \mapsto x^3 + x(\mathbf{R} \to \mathbf{R})$. The value $f(x)$ of the function at x is also called the *image of x under the mapping f*.

If we are given two mappings $f : X \to Y$ and $g : Y \to Z$ then we can combine them to obtain a mapping from X to Z in an obvious way. Starting with an element x of X we first map it to its image $f(x)$ under f and then map this element of Y to its image $g(f(x))$ under g. The mapping thus defined is called the *composite* of f and g and is denoted by $g \circ f$. It is given by the formula $(g \circ f)(x) = g(f(x))$ for all $x \in X$. The reader will have met composite functions before under the name "function of a function". Composition of functions is the same thing as substitution of one function in another. The composite function $g \circ f$ exists only when the domain of g is the same as the range of f.

Example 2.8. The function $x \mapsto \sin^2 x$ $(\mathbf{R} \to \mathbf{R})$ is the composite $g \circ f$ of the function $f : x \mapsto \sin x$ and the function $g : y \mapsto y^2$ (both functions from $\mathbf{R}$ to $\mathbf{R}$). Note that in this special situation the functions can be composed the other way round to give the function $f \circ g : \mathbf{R} \to \mathbf{R}$. This function sends x to $\sin(x^2)$ and is not the same function as $g \circ f$.

THEOREM 2A. *Composition of functions is associative. More precisely, if $f : A \to B$, $g : B \to C$ and $h : C \to D$, then the two functions $h \circ (g \circ f)$ and $(h \circ g) \circ f$ from A to D are equal.*

Proof. If k_1 denotes the function $h \circ (g \circ f)$ and k_2 denotes $(h \circ g) \circ f$ then it follows from the definition of composition that for each $a \in A$, $k_1(a) = h((g \circ f)(a)) = h(g(f(a)))$ and similarly $k_2(a) = (h \circ g)(f(a)) = h(g(f(a)))$. Thus $k_1(a) = k_2(a)$ for all $a \in A$, that is, $k_1 = k_2$.

Given a set A there is a special function from A to A called the *identity function* on A. It maps each element of A to itself and it is denoted by ι_A. Thus $\iota_A(a) = a$ for all $a \in A$. It is clear that the identity functions play a special role with respect to composition; if $f: X \to Y$ then $f \circ \iota_X = f$ and $\iota_Y \circ f = f$. They behave rather like the 1 in elementary algebra, but in a one-sided way (see law (M2) in Example 1.1). This fact leads us to define *inverse functions* as follows. Let $f: X \to Y$ and $g: Y \to X$; then the two composite functions $f \circ g: Y \to Y$ and $g \circ f: X \to X$ both exist. If $f \circ g = \iota_Y$ and $g \circ f = \iota_X$ we say that the functions f and g are inverse to each other. In other words, f and g are inverse functions if $f(g(y)) = y$ and $g(f(x)) = x$ for all $x \in X$ and all $y \in Y$. Not all functions have inverses and we shall shortly prove an easy criterion for a function to have an inverse. We postpone examples of inverses until then.

If a function $f: X \to Y$ has the property that no two distinct elements of X have the same image under f then f is called an *injection* and we say that f embeds X in Y. Thus f is an injection if and only if the statement $f(x_1) = f(x_2) \Rightarrow x_1 = x_2$ is true. (Here we have omitted the quantifiers $\forall x_1 \in X$ and $\forall x_2 \in X$ which may be taken for granted, as is done in normal English. Only when there is danger of misunderstanding will we attach such quantifiers to statements of implication.)

Example 2.9. The function $x \mapsto x^2$ ($\mathbf{R} \to \mathbf{R}$) is not an injection because $2 \mapsto 4$ and $-2 \mapsto 4$. The function $x \mapsto x^3$ ($\mathbf{R} \to \mathbf{R}$) *is* an injection, because for real numbers x_1, x_2 we have $x_1^3 = x_2^3 \Rightarrow x_1 = x_2$. However, the function $x \mapsto x^3$ ($\mathbf{C} \to \mathbf{C}$) is *not* an injection because there are three distinct cube-roots of 1 in $\mathbf{C}$.

Example 2.10. If A is a subset of B then the function $A \to B$ defined by $a \mapsto a$ for all $a \in A$, is an injection. It is called the inclusion map of A in B. (Note: this is not the same function as the identity function ι_A since the identity function has range A while the inclusion function has range B.)

A mapping $f: X \to Y$ is called a *surjection* if every member of Y is the image of at least one member of X, that is, if $(\forall y \in Y)(\exists x \in X)(f(x) = y)$. A mapping which is both an injection and a surjection is called a *bijection* or a *one-one correspondence*.

For convenience in giving examples, we introduce standard notations for intervals on the real line. If $a, b \in \mathbf{R}$ and $a \leqslant b$ then $[a, b]$ denotes the closed interval $\{x \in \mathbf{R}; a \leqslant x \leqslant b\}$ which includes the end-points a and b. If we omit the end-points we get the open interval

$\{x \in \mathbf{R};\ a < x < b\}$, denoted by (a,b). We also write

$[a,b)$ for $\{x \in \mathbf{R}; a \leqslant x < b\}$,
$[a,\infty)$ for $\{x \in \mathbf{R}; x \geqslant a\}$,
$(-\infty, b)$ for $\{x \in \mathbf{R}; x < b\}$,

with similar conventions for $(a,b]$, (a,∞), $(-\infty, b]$.

Example 2.11. The map $x \mapsto x^2$ from $\mathbf{R}$ to $[0, \infty)$ is a surjection but not an injection. The map $x \mapsto x^2$ from $[0, \infty)$ to $\mathbf{R}$ is an injection but not a surjection. The map $x \mapsto x^2$ from $[0, \infty)$ to $[0, \infty)$ is a bijection. So is the map $x \mapsto x^3$ from $\mathbf{R}$ to $\mathbf{R}$.

Example 2.12. If we consider surfaces as being sets of points then geographical maps become maps in our sense. For example, if X denotes the surface of the earth (idealised as a sphere) with its two poles removed and if Y denotes a cylinder on the equator with axis along the polar axis, then Mercator's projection (i.e. projection from the centre of the earth) gives a function $f: X \to Y$ which is in fact a bijection.

THEOREM 2B. *Let $f: X \to Y$ and $g: Y \to Z$. Then*
 (i) *if f and g are both injections then $g \circ f: X \to Z$ is an injection;*
 (ii) *if f and g are both surjections then $g \circ f$ is a surjection;*
 (iii) *if f and g are both bijections then $g \circ f$ is a bijection.*

Proof. (i) Let $x_1, x_2 \in X$ and suppose $x_1 \neq x_2$. If f is an injection then $f(x_1) \neq f(x_2)$. If g is also an injection, it follows that $g(f(x_1)) \neq g(f(x_2))$. Thus $(g \circ f)(x_1) \neq (g \circ f)(x_2)$ whenever $x_1 \neq x_2$; in other words $g \circ f$ is an injection.

(ii) Let $z \in Z$. If g is a surjection then $z = g(y)$ for some $y \in Y$. If f is also a surjection, this element y can be written $y = f(x)$ for some $x \in X$. Hence $z = g(y) = g(f(x)) = (g \circ f)(x)$ for some x. Since this is true for every $z \in Z$, $g \circ f$ is a surjection.

(iii) This follows immediately from (i) and (ii).

THEOREM 2C. *A function $f: X \to Y$ has an inverse $g: Y \to X$ if and only if f is a bijection. The inverse function g, if it exists, is uniquely determined by f and is also a bijection.*

Proof. Suppose first that f has an inverse g. Then $g(f(x)) = x$ for all $x \in X$, and $f(g(y)) = y$ for all $y \in Y$. If $x_1, x_2 \in X$ and $f(x_1) = f(x_2)$, it follows that $x_1 = g(f(x_1)) = g(f(x_2)) = x_2$; hence f is an injection. Also, if $y \in Y$, then $y = f(g(y))$ is the image under f of the element $g(y)$ of X; hence f is a surjection. This shows that f is a bijection. Also, since g has an inverse (namely f) the same argument shows that g is a bijection. The uniqueness of g follows easily; for suppose g_1 and g_2 are two functions inverse to f. Then $f(g_1(y)) = f(g_2(y)) = y$ for all $y \in Y$. Since f is an injection, this implies that $g_1(y) = g_2(y)$ for all $y \in Y$, i.e. $g_1 = g_2$.

Now suppose that f is a bijection. Then, for every $y \in Y$, there is at least one $x \in X$ such that $f(x) = y$ (because f is a surjection). But, given y, there cannot be more than one $x \in X$ such that $f(x) = y$ (because f is an injection). We may therefore define $g(y)$ to be that unique $x \in X$ such that $f(x) = y$, and we obtain in this way a function $g: Y \to X$. It is clear that g is inverse to f since (i) from the construction of g we have $f(g(y)) = f(x) = y$ for every $y \in Y$ and (ii) if $x \in X$ and if we write y for $f(x)$ then $g(y)$ is, by construction, equal to x, so $g(f(x)) = x$.

Example 2.13. (i) The function $x \mapsto x^2$ ($[0, \infty) \to [0, \infty)$) is a bijection and its inverse is $y \mapsto \sqrt{y}$ ($[0, \infty) \to [0, \infty)$).

(ii) The function $x \mapsto 2x^3 - 1$ ($\mathbf{R} \to \mathbf{R}$) is a bijection since it is the composite of the three bijections $x \mapsto x^3$, $y \mapsto 2y$ and $z \mapsto z - 1$. Its inverse is the function $t \mapsto \sqrt[3]{\frac{1}{2}(t + 1)}$.

Example 2.14. The exponential series

$$\sum_{n=0}^{\infty} x^n/n!$$

converges for every real number x and its sum, denoted by e^x, is a real number uniquely determined by x. We therefore have a function $x \mapsto e^x$ ($\mathbf{R} \to \mathbf{R}$), called the exponential function. From elementary analysis we know the following facts about the exponential function: (i) it is a strictly increasing function; (ii) its values are always positive; (iii) it is continuous; and

(iv) $\lim_{x \to -\infty} e^x = 0$, $\lim_{x \to +\infty} e^x = +\infty$.

From (i) it follows that the exponential function does not take the same value twice, i.e. $x \mapsto e^x$ ($\mathbf{R} \to \mathbf{R}$) is an injection. It is clearly not a surjection, because of (ii). However, we may consider the restricted function $x \mapsto e^x$ ($\mathbf{R} \to (0, \infty)$), and this *is* a surjection, as follows from (iii) and (iv). The argument is that by (iv) e^x takes arbitrarily small and arbitrarily large positive values and, being continuous, it takes all values in between, hence takes *all* positive values. (See, for example, Burkhill, *A First Course in Mathematical Analysis* (Cambridge University Press, 1962), pp. 105–107). Thus we have a bijection $x \mapsto e^x$ from $\mathbf{R}$ to $(0, \infty)$ and it has a unique inverse from $(0, \infty)$ to $\mathbf{R}$. This inverse function is called the logarithmic function and is denoted by $y \mapsto \log y$. It is defined for all $y > 0$ by the rule: $\log y$ is the unique real number x such that $e^x = y$; and being inverse to the exponential function, it satisfies $e^{\log y} = y$ for all $y > 0$ and $\log(e^x) = x$ for all $x \in \mathbf{R}$.

This last example shows that bijections and their inverses are important in analysis. Their importance in set theory stems from the process of counting. To count a set of objects and to decide that there are n objects is to construct a one-one correspondence (bijection) from the given set to the set $\{1, 2, 3, \ldots, n\}$. We say that the set X is *similar* to the set Y if there exists a bijection from X to Y, and we write $X \simeq Y$ if this is the case. By Theorem 2C, if $X \simeq Y$, then $Y \simeq X$. Also, by Theorem 2B, if $X \simeq Y$ and $Y \simeq Z$ then $X \simeq Z$. A set X is *finite* if it is similar to the set $\{1, 2, \ldots, n\}$ for some $n \in \mathbf{N}$. This number n is called the *cardinal* of X or the *order* of X and is denoted by $|X|$. It is uniquely determined by X since no two of the sets $\{1, 2, \ldots, n\}$ for different n are similar. (We shall justify this last statement in Chapter 3.) A set which is not finite is called *infinite*. For example, $\mathbf{N}$ itself is infinite. A set which is similar to $\mathbf{N}$ or to a subset of $\mathbf{N}$ is called *countable*.

Example 2.15. All finite sets are countable. The set of all integers is countable by means of the function $f : \mathbf{Z} \to \mathbf{N}$ defined by

$$\begin{cases} f(x) = 2x & \text{if } x \geqslant 0, \\ f(x) = -2x - 1 & \text{if } x < 0. \end{cases}$$

This counts the integers in the order $0, -1, 1, -2, 2, -3, \ldots$. The set of all rational numbers is also countable, a somewhat surprising fact which may be seen in many ways. For example, every non-zero rational number x is uniquely expressible in the form $x = (-1)^r p/q$, where p and q are positive integers with highest common factor 1 and r is 1 or 2. If we map x to $2^r 3^p 5^q$ and map 0 to 0 we obtain an *injection* $\mathbf{Q} \to \mathbf{N}$ which establishes the similarity of $\mathbf{Q}$ with a subset of $\mathbf{N}$. With more effort one can find bijection $\mathbf{Q} \to \mathbf{N}$.

The set $\mathbf{R}$ of all real numbers is *not* countable. Thus it is not true that all infinite sets are similar, and this fact leads to the study of infinite cardinal numbers, a subject outside the scope of this book. See, for example, Halmos, *Naive Set Theory* (Van Nostrand, 1960), pp. 90–98.

The definition of function given above seems to exclude the possibility of functions of several variables. However a simple device enables us to discuss several variables as if they were one. Given two sets A, B we define their *product* $A \times B$ to be the set whose members are all possible ordered pairs* (a, b) with $a \in A$ and $b \in B$. Here an ordered pair means something more special than a set with two elements. In the set $\{a, b\}$ the order in which the elements are written is irrelevant: $\{a, b\} = \{b, a\}$. However, for ordered pairs, $(a, b) \neq (b, a)$ unless $a = b$; in fact $(a, b) = (c, d)$ if and only if $a = c$ and $b = d$. Thus, if $|A| = m$ and $|B| = n$, there are n different pairs (a, b) for each of the m

---

*It is unfortunate that the notation (a, b) is also used for open intervals in $\mathbf{R}$, but the context will always show which meaning is intended.

possible choices of $a \in A$, so $|A \times B| = mn$. The product $A \times B \times C$ of three sets is the set of ordered triples (a, b, c) where $a \in A$, $b \in B$, $c \in C$ and there is no danger in identifying this product with $(A \times B) \times C$ or with $A \times (B \times C)$, these different products being similar by means of obvious maps in which the elements (a, b, c), $((a, b), c)$ and $(a, (b, c))$ correspond. The same applies to all products of finitely many sets $A_1 \times A_2 \times \ldots \times A_r$.

Consider now a function $f : X \times Y \to Z$, where X, Y, Z are arbitrary sets. Given an element $(x, y) \in X \times Y$, the function f determines an element of Z, denoted by $f(x, y)$. The value of f is determined uniquely by the element $x \in X$ together with the element $y \in Y$. This is what we mean by a function of two variables, and it should be noted that the two variables may be of quite different types, coming from different sets X, Y.

Example 2.16. The set $\mathbf{R} \times \mathbf{R}$ (also written $\mathbf{R}^2$) consists of all ordered pairs (x, y) of real numbers. This set is known as the Cartesian plane or the real plane and is the object of study in coordinate geometry. The elements (x, y) are called *points* and the numbers x and y are called the coordinates of the point (x, y). (There are other, more geometric, ways of defining planes, but one is then faced with the task of proving that they can be coordinatized in this way.) A function $f : \mathbf{R} \times \mathbf{R} \to \mathbf{R}$ such as $(x, y) \mapsto x^2 + y^2$ is just a real-valued function of two real variables in the traditional sense, and it can be thought of geometrically as a mapping from the real plane to the real line. On the other hand a function $g : \mathbf{R} \to \mathbf{R} \times \mathbf{R}$ such as $t \mapsto (\cos t, \sin t)$ is known as a parametric curve; this particular one maps the real line into the plane so that its points all go to points of a circle. Functions from the plane to the plane such as $(x, y) \mapsto (x \cos y, x \sin y)$ also appear frequently in analysis. These are all now special cases of the one definition of function.

A function $f : A \times A \to A$ is called a *binary operation on* A. It can be applied to any two elements a_1, a_2 of A to produce a third element $f(a_1, a_2)$. We can define in the same way ternary operations $A^3 = A \times A \times A \to A$ and, in general, n-ary operations $A^n \to A$. In particular, a *unary operation on* A is just a function from A to A.

The study of operations on sets is a major part of algebra and this book is chiefly devoted to the most frequently met unary and binary operations. Ternary and higher operations are of more academic interest and will not be mentioned again. Binary operations often have special symbols such as $+, \times, \circ$ which are placed *between* the variables. Unary operations have a multitude of different notations.

Example 2.17. Addition is a binary operation on $\mathbf{R}$ given by $(x, y) \mapsto x + y$. The same formula defines a binary operation on $\mathbf{Z}$ since the sum of two integers is again an integer. Similarly $x \mapsto -x$ is a unary operation on $\mathbf{R}$ or on $\mathbf{Z}$, but not of course on $\mathbf{N}$. Note that inversion

$x \mapsto x^{-1}$ is not an operation on **R** since it cannot be applied to $x = 0$. However $x \mapsto x^{-1}$ *is* a unary operation on $(0, \infty)$.

Example 2.18. It is important, when specifying an operation, to make sure that its domain is stated and also that it is a properly defined function. The commonest pitfall is exemplified in the following. Every rational number q can be written a/b where $a, b \in \mathbf{Z}$ and $b \neq 0$. It would seem feasible therefore to define a binary operation $*$ on **Q** by a rule such as

$$\frac{a}{b} * \frac{c}{d} = \frac{a+c}{bd}.$$

However the rational number q does not determine its numerator a and denominator b uniquely, so there is no reason to expect that $(a+c)/bd$ will be uniquely determined by the rational numbers a/b and c/d. In fact it is not uniquely determined because, using the given definition, we find

$$\frac{1}{2} * \frac{1}{3} = \frac{2}{6} = \frac{1}{3}.$$

But

$$\frac{2}{4} * \frac{1}{3} = \frac{3}{12} \neq \frac{1}{3},$$

even though

$$\frac{1}{2} = \frac{2}{4}.$$

On the other hand the formula

$$\frac{a}{b} * \frac{c}{d} = \frac{ac}{bd}$$

does define a binary operation on **Q** (multiplication of rational numbers) because, if

$$\frac{a}{b} = \frac{a'}{b'} \quad \text{and} \quad \frac{c}{d} = \frac{c'}{d'},$$

one can deduce that

$$\frac{ac}{bd} = \frac{a'c'}{b'd'}.$$

We shall return to this point in later chapters but the reader is advised to think hard about the difference between these two formulae, and to do Exercise 16.

We end our survey of elementary set theory with a brief look at correspondences and relations. We shall make considerable use later of equivalence relations and, to a lesser extent, of order relations.

First, a *correspondence* from a set X to a set Y is given by a rule of a more general type than a functional rule. The rule specifies that certain elements of X "correspond" to certain elements of Y and the only requirement is that, for each pair $(x, y) \in X \times Y$ the rule should determine whether or not x corresponds to y. If γ denotes such a correspondence, we write $x \gamma y$ to mean that x corresponds to y in the correspondence γ. In this notation γ is used as a verb, and $x \gamma y$ is a statement. Examples of symbols for correspondences used in this way are $=, >, \leq, \neq, \in, \subset, \supset, \perp$. A function from X to Y is a special kind of correspondence γ from X to Y namely, one satisfying the conditions:
(i) for every $x \in X$ there is at least one $y \in Y$ such that $x \gamma y$, and
(ii) for all $x \in X$ and all $y_1, y_2 \in Y$, if $x \gamma y_1$ and $x \gamma y_2$ then $y_1 = y_2$.

Example 2.19. Let X be the set of all points in the plane and let Y be the set of all lines in the plane. Then incidence between points and lines is a correspondence γ from X to Y defined by $x \gamma y \Leftrightarrow$ the point x is on the line y. Equally, of course, the statement "the point x is not on the line y" defines a correspondence, as does *any* statement about x and y which is meaningful and either true or false for each possible pair (x, y). Two correspondences γ, γ' from X to Y are to be considered equal if $x \gamma y \Leftrightarrow x \gamma' y$.

For any set X, a correspondence from X to itself is called a *relation on X*. For example, if X is the set of humans, then the concept "uncle" defines a relation ρ on X by the rule $a \rho b \Leftrightarrow a$ is the uncle of b, where $a, b \in X$. Similarly, "brother", "father", "ancestor", "acquaintance", "enemy" all define relations on X. We do not have to look far for mathematical examples: $\geq$ on **R**; $=$ on any set; $\subset$ on the set of subsets of a given set; divisibility on **Z** ($m \rho n$ if n is an integer multiple of m). Also, any equation $f(x, y) = 0$ defines a relation on **R** if f is a function from $\mathbf{R}^2$ to **R**; in this relation ρ, $x \rho y \Leftrightarrow f(x, y) = 0$.

Various types of relations with special properties are frequently met in mathematics. We single out the following types:

(i) a relation ρ on X is *reflexive* if $x \rho x$ for every $x \in X$;
(ii) a relation ρ on X is *symmetric* if, for all $x, y \in X$, $x \rho y \Rightarrow y \rho x$;
(iii) a relation ρ on X is *antisymmetric* if, for all $x, y \in X$, $(x \rho y$ and $y \rho x) \Rightarrow x = y$;
(iv) a relation ρ on X is *transitive* if, for all $x, y, z \in X$, $(x \rho y$ and $y \rho z) \Rightarrow x \rho z$.

Example 2.20. On the set of all humans, the relation "brother" is not symmetric, but the relation "sibling" is symmetric; the relation "parent" is not transitive, but the relation "ancestor" is transitive. The relation $\geqslant$ on $\mathbf{R}$ is reflexive, antisymmetric and transitive. The relation $\neq$ on $\mathbf{R}$ is symmetric (but not transitive). The relation $=$ on $\mathbf{R}$ is reflexive, symmetric, antisymmetric and transitive.

The main uses of relations in mathematics are for purposes of *comparison* and *classification*. For the comparison or ranking of the elements of a set X, one needs to introduce a relation ρ on X for which $x \rho y$ means that x dominates or precedes y in some respect. Such relations will usually be *transitive* and *antisymmetric,* and for convenience one usually defines them so as to be *reflexive* as well. A relation having these three properties is called an *order relation*, and familiar examples are: $\geqslant$ on $\mathbf{R}$; $\leqslant$ on $\mathbf{R}$; $\subset$ on the set of subsets of a fixed set; divisibility on the set of positive integers (but not divisibility on $\mathbf{Z}$ because 1 and -1 both divide each other). The relation $\geqslant$ on $\mathbf{R}$ is a particularly nice order relation because *any two real numbers can be compared* by it: either $x \geqslant y$ or $y \geqslant x$. Such a relation (i.e. an order relation ρ on X such that $(\forall x, y \in X)(x \rho y \text{ or } y \rho x))$ is called a *total order relation* or a *linear order relation* because it provides a complete ranking of the elements of X. The relations $\subset$ on the set of subsets of a set and divisibility on the set of positive integers are *not* linear order relations.

For purposes of classification a rather different type of relation is needed. To classify the elements of a set X is to divide the set into disjoint subsets, i.e. to form a partition of X. In precise terms a *partition of X* is a set of subsets X_i of X, where i varies in some indexing set I, such that

(i) $\bigcup_{i \in I} X_i = X$ and

(ii) $X_i \cap X_j = \emptyset$ if $i \neq j$.

The sets X_i will be called the classes of the partition. Given a relation ρ on X we often would like to classify the elements of X so that two elements are in the same class if and only if they are related by ρ. Sometimes this is possible and sometimes not. For example if X is the set of humans and $x \rho y$ means that x and y have the same number of hairs on their heads then the classification is possible. If, on the other hand, $x \rho y$ means that x is acquainted with y then such a classification is not possible. These are obvious examples, but if ρ is a mathematical relation such as the relation on $\mathbf{Z}$ in which $x \rho y$ means that $x - y$ is divisible by 7, it may not be so obvious whether a classification is possible. We therefore need a criterion for deciding whether a given relation is suitable for classification, and this is given by the following definition and theorem.

Definition. A relation ρ on a set X is an *equivalence relation* if it is reflexive, symmetric and transitive.

THEOREM 2D. (i) Given a partition $X = \bigcup_{i \in I} X_i$ of the set X, the relation σ on X defined by "$x \sigma y \Leftrightarrow x$ and y are in the same class X_i of the partition" is an equivalence relation.

(ii) If ρ is any equivalence relation on X then there is a partition $X = \bigcup_{i \in I} X_i$ of X such that the elements x and y of X are in the same class X_i if and only if $x \rho y$.

Proof. (i) It is obvious from the definition of partition that the relation σ defined in this way is an equivalence relation.

(ii) Suppose that ρ is an equivalence relation on X. For each $x \in X$ define $\langle x \rangle = \{y \in X; y \rho x\}$. Then $x \in \langle x \rangle$ (because $x \rho x$) and we call $\langle x \rangle$ the *equivalence class containing* x. We shall prove that

(a)   if $x \rho y$ then $\langle x \rangle = \langle y \rangle$, and
(b)   if $\neg\, x \rho y$ then $\langle x \rangle \cap \langle y \rangle = \emptyset$.

First, if $x \rho y$ and if t is any member of $\langle x \rangle$, then $t \rho x$ by definition and hence $t \rho y$, because ρ is transitive. Thus $t \in \langle y \rangle$ and we have shown that $\langle x \rangle \subset \langle y \rangle$ whenever $x \rho y$. But ρ is symmetric, so if $x \rho y$ then $y \rho x$, which implies $\langle y \rangle \subset \langle x \rangle$ by the same argument. This proves (a). On the other hand, if $\langle x \rangle$ and $\langle y \rangle$ have an element in common, say $u \rho x$ and $u \rho y$, then $x \rho u$ by symmetry and so $x \rho y$ by transitivity. This proves (b). The statements (a) and (b) together imply that distinct equivalence classes are disjoint and that two elements x and y are in the same class if and only if $x \rho y$. The union of all the classes $\langle x \rangle$ is X (since $x \in \langle x \rangle$) so the collection of all distinct sets of the form $\langle x \rangle$ is a partition of X with the required properties.

Example 2.21. On $\mathbf{R}^2$, the relation $\sim$ defined by $(x, y) \sim (x', y') \Leftrightarrow y - y' = 2(x - x')$ is an equivalence relation. The equivalence classes are the parallel lines $y = 2x + c$ for various real numbers c.

Example 2.22. Let A be the set $\mathbf{R}^2$ with the origin $(0, 0)$ removed. The relation $\sim$ on A defined by

$$(x, y) \sim (x', y') \Leftrightarrow (\exists \lambda \in \mathbf{R})(\lambda x = x' \text{ and } \lambda y = y')$$

is an equivalence relation. A typical equivalence class consists of all points on a line through the (missing) origin.

Example 2.23. Let $f : X \to Y$ be any function. Then the relation $\sim$ on X defined by

$$x_1 \sim x_2 \Leftrightarrow f(x_1) = f(x_1)$$

is an equivalence relation and its equivalence classes are called the *fibres* of f. A typical fibre is the set of all $x \in X$ such that $f(x)$ takes a particular value.

This last example is, in a sense, typical. Any equivalence relation can be related to a function in the way described. Given an equivalence relation ρ on X we construct a new set X/ρ whose *members* are the equivalence classes of ρ. Thus X/ρ is a set of subsets of X and must not be confused with X itself. If $x \in X$ then the equivalence class $\langle x \rangle$ containing x is a member of X/ρ, and we can define a mapping f from X to X/ρ by the rule $x \mapsto \langle x \rangle$ for every $x \in X$. Then $f(x_1) = f(x_2) \Leftrightarrow \langle x_1 \rangle = \langle x_2 \rangle \Leftrightarrow x_1 \rho x_2$. So the fibres of this mapping are the equivalence classes of ρ. The set X/ρ is called the *quotient set of X by the equivalence relation ρ*, and the mapping $x \mapsto \langle x \rangle$ is called the *canonical mapping* or the *quotient mapping* from X to X/ρ. The construction of quotient sets is a very important method of producing new mathematical objects from old ones, and we shall see its use later in several contexts.

Example 2.24. In Example 2.21 the quotient set $\mathbf{R}^2/\sim$ is the set of all lines of slope 2 in the plane $\mathbf{R}^2$, and the quotient mapping sends each point to the line through it with slope 2. In Example 2.22 the quotient set $A/\sim$ is the set of all lines through the origin in $\mathbf{R}^2$, with the origin removed. There is one element in $A/\sim$ for each possible slope, i.e. one for each real number and one of infinite slope. So $A/\sim$ is similar to $\mathbf{R}$ with an extra element ∞ adjoined. It is called the real projective line and is a basic object of study in projective geometry, in which the points of a line are a represented by pairs (x, y) of real numbers, not both zero, but in which (x, y) and $(\lambda x, \lambda y)$ always represent the same point.

THEOREM 2E. *Let ρ be an equivalence relation on the set A and let $q : A \to A/\rho$ be the quotient map. Let $f : A \to B$ be a given function. Then the following two conditions are equivalent (i.e., each implies the other):*

(i)   *there exists a function $g : A/\rho \to B$ such that $g \circ q = f$;*
(ii) *for all $x, y \in A$, $x \rho y \Rightarrow f(x) = f(y)$.*

Proof. (i) $\Rightarrow$ (ii): By definition of q, if $x \rho y$ then $q(x) = q(y) = $ the equivalence class containing x and y. If $f = g \circ q$ for some $g : A/\rho \to B$ it follows that $f(x) = g(q(x)) = g(q(y)) = f(y)$.

(ii) $\Rightarrow$ (i): Suppose that condition (ii) is satisfied. Then f is *constant* on each equivalence class, that is, it takes the same value at all members of a given equivalence class. Hence, for each equivalence class $X \in A/\rho$ we can define $g(X) = f(x)$, where x is an arbitrary member of X. This $g(X)$ is uniquely determined by X, because of the constancy of f on X, so we have defined a function $g : A/\rho \to B$. Since $(g \circ q)(x) = g(\langle x \rangle) = f(x)$ for all $x \in A$, we have $g \circ q = f$.

The property of the quotient map proved in this theorem is the first example of a "universal" property. It follows from the theorem that $q : A \to A/\rho$ is *universal* amongst functions f with domain A such that

$x \, \rho \, y \Rightarrow f(x) = f(y)$. This means that (a) $x \, \rho \, y \Rightarrow q(x) = q(y)$ and (b) every f satisfying $x \, \rho \, y \Rightarrow f(x) = f(y)$ can be obtained by composing q with another function. Other examples of universal properties will occur later.

Exercises

1. Prove that if A and B are subsets of a set S, then
$A \supset B \Leftrightarrow A \cap B = B \Leftrightarrow A \cup B = A$.

2. Verify the distributive laws $A \cup (B \cap C) = (A \cup B) \cap (A \cup C)$ and $A \cap (B \cup C) = (A \cap B) \cup (A \cap C)$ and the de Morgan laws $(A \cup B)' = A' \cap B'$ and $(A \cap B)' = A' \cup B'$ for subsets A, B, C of a set S.

3. Which of the following are functions between the stated sets? Of those which are functions, which are injections, surjections, bijections? Find the inverse functions of all the bijections.

   (i) $f(x) = 1 - x^2$ ($\mathbf{R} \to \mathbf{R}$);

   (ii) $f(x) = 1 - x^2$ ($[-1, 1] \to [0, 1]$);

   (iii) $f(x) = (1 - x)/(1 + x)$ $((-1, \infty) \to (-1, \infty))$;

   (iv) $f(x) = x^n$ ($\mathbf{R} \to \mathbf{R}$), where n is a positive integer;

   (v) $f(x) = \begin{cases} x + 1 \text{ if } x \text{ is even} \\ x - 1 \text{ if } x \text{ is odd} \end{cases}$ ($\mathbf{Z} \to \mathbf{Z}$);

   (vi) $f(x) = \begin{cases} x^3 \text{ if } x \text{ is rational} \\ x \;\; \text{ if } x \text{ is irrational} \end{cases}$ ($\mathbf{R} \to \mathbf{R}$);

   (vii) $f(x) = \tan x$ ($[0, \pi] \to \mathbf{R}$);

   (viii) $f(x) = \tan x \left(\left(-\frac{\pi}{2}, \frac{\pi}{2} \right) \to \mathbf{R} \right)$.

4. If A and B are finite sets with $|A| = m$, $|B| = n$, how many different injections are there from A to B?
(Harder) Prove that the number of surjections from A to B is

$$\sum_{i=0}^{n} (-1)^i \binom{n}{i} (n - i)^m.$$

(Here $\binom{n}{i}$ is, as usual, the binomial coefficient $n!/i!(n-i)!$. The number of surjections is closely related to Stirling numbers; see Riordan, *Introduction to Combinatorial Analysis* (Wiley, 1958), p. 91.)

5. Prove that if $f : A \to B$ is an injection then, for any function $h : A \to C$ there exists a function $g : B \to C$ such that $g \circ f = h$. Verify that the function $f : \mathbf{R} \to \mathbf{R}^2$ defined by $f(t) = ((t + 1)^2, (2t + 1)^2)$ is an injection, and find a function $g : \mathbf{R}^2 \to \mathbf{R}$ such that $g \circ f$ is the identity function on $\mathbf{R}$.

6. Let $f : A \to B$ and $h : A \to C$ be given functions. Prove that there exists a function $g : B \to C$ such that $g \circ f = h$, if and only if h is constant on the fibres of f (i.e., $f(a_1) = f(a_2) \Rightarrow h(a_1) = h(a_2)$).

7. Let $f : \mathbf{R} \to \mathbf{R}$ be any function satisfying $f(-x) = f(x)$ for all $x \in \mathbf{R}$. Prove that there is a unique function $g : [0, \infty) \to \mathbf{R}$ such that $f(x) = g(x^2)$ for all $x \in \mathbf{R}$.

8. Let $f : \mathbf{R} \to \mathbf{R}^2$ be the function defined by $f(x) = (\cos x, \sin x)$. What are the fibres of f? Prove that if $g : \mathbf{R} \to \mathbf{R}$ is any function which is periodic with period 2π (i.e., $g(x + 2\pi) = g(x)$ for all $x \in R$) then g can be written in the form $g(x) = h(\cos x, \sin x)$, where h is a suitable function of two variables from $\mathbf{R}^2$ to $\mathbf{R}$.

9. Let $f : A \to X$ and $g : A \to Y$ be given functions and let h be the function from A to $X \times Y$ defined by $h(a) = (f(a), g(a))$. Prove that h is a bijection if and only if:

   (i) f and g are both surjections, and
   (ii) each fibre of f has exactly one member in common with each fibre of g.

10. Which of the following relations ρ are equivalence relations; which are order relations? When ρ is an equivalence relation, describe the equivalence classes.

    (i)   In $\mathbf{Z}$; $x \rho y$ means "x divides y and y divides x";
    (ii)   In $\mathbf{R}$; $x \rho y$ means "$x - y$ is rational";
    (iii) In $\mathbf{R}$; $x \rho y$ means "$xy \geq 0$";
    (iv)   In $\mathbf{Z} \times \mathbf{Z}$; $(a, b) \rho (c, d)$ means "$ad = bc$";
    (v)   In $\mathbf{Z} \times \mathbf{Z}'$, where $\mathbf{Z}'$ is the set of non-zero integers; ρ is as in (iv);
    (vi)   In $\mathbf{R}$; $x \rho y$ means "$(\forall n \in \mathbf{Z})(n < x \Leftrightarrow n < y)$";
    (vii) In $\mathbf{R}^2$; $(x, y) \rho (x', y')$ means "$(\exists \lambda \in \mathbf{R})(\lambda > 0$ and $x' = \lambda x$ and $y' = \lambda y)$";
    (viii) In $\mathbf{R}^2$; $(x, y) \rho (x', y')$ means "$(\exists \lambda \in \mathbf{R})(\lambda \geq 1$ and $x' = \lambda x$ and $y' = \lambda y)$".

11. How many different equivalence relations are there on a set with four elements?

12. Resolve the following paradox.
    Proposition. Any symmetric transitive relation ρ is reflexive.
    Proof. Let $a \rho b$. Then $b \rho a$ by symmetry. But

30 SET THEORY

$(a \rho b$ and $b \rho a) \Rightarrow a \rho a$ by transitivity. Hence $a \rho a$ for all a.
Counter-example: In **R**, let $x \rho y$ mean $xy > 0$.

13. Call a relation ρ on a set A *circular* if $(\forall x, y, z \in A)$
$(x \rho y \,\&\, y \rho z \Rightarrow z \rho x)$. Prove that any reflexive circular relation is an equivalence relation.

14. Let $A = A_1 \times A_2$ and let ρ_1, ρ_2 be equivalence relations on A_1, A_2 respectively. Let ρ be the relation on A defined by $(a_1, a_2) \rho (b_1, b_2) \Leftrightarrow a_1 \rho_1 b_1$ and $a_2 \rho_2 b_2$.
Prove that ρ is an equivalence relation and describe its equivalence classes. Prove that A/ρ is similar to $(A_1/\rho_1) \times (A_2/\rho_2)$.

15. Let ρ be a relation on A which is reflexive and transitive. Let $x \sigma y$ mean "$x \rho y$ and $y \rho x$". Prove that σ is an equivalence relation.
Let $B = A/\sigma$ and define a relation $\bar{\rho}$ on B by the rule:

$$X \bar{\rho} Y \Leftrightarrow (\forall x \in X)(\forall y \in Y)(x \rho y),$$

where X, Y stand for equivalence classes of σ. Prove that $\bar{\rho}$ is an order relation on B.
[Examples: (i) In **R**, $x \rho y$ means "$\sin x \leqslant \sin y$";
(ii) in **Z**, $x \rho y$ means "x divides y". In each of these examples describe σ, B and $\bar{\rho}$.]

16. Which of the following formulae define binary operations $*$ on **Q**? In each case m, n, m', n' stand for integers with $n \neq 0, n' \neq 0$.

(i) $\dfrac{m}{n} * \dfrac{m'}{n'} = \dfrac{mn' + m'n}{nn'};$

(ii) $\dfrac{m}{n} * \dfrac{m'}{n'} = \dfrac{mn' - m'n}{mn' + m'n};$

(iii) $\dfrac{m}{n} * \dfrac{m'}{n'} = \dfrac{m^2 n' + m'^2 n}{(nn')^2}.$

CHAPTER 3

The Integers

The set **Z** of integers can be characterized in many ways; we choose a description in terms of the operations $+, -, \cdot$, the constants 0, 1 and the relation $\leq$. This description is concerned only with the algebraic relations which hold between integers. It tells us nothing of the nature of the individual integers themselves, a subject which is of more philosophical than mathematical interest.

The basic facts about **Z** that we shall assume are the following.

I. The binary operations $+, \cdot$ on **Z**, the unary operation $-$ on **Z** and the constants $0, 1 \in \mathbf{Z}$ satisfy all the laws of standard algebra except (M3). (See Chapter 1, Example 1.1). In place of (M3) the integers satisfy the *cancellation law:*

(M3′)   if $xy = xz$ and $x \neq 0$, then $y = z$.

II. The relation $\leq$ on **Z** is a linear order relation (recall from Chapter 2 that this means a reflexive, transitive, anti-symmetric relation such that for any two integers, x, y either $x \leq y$ or $y \leq x$). It is related to the operations on **Z** by the laws:

(O1)   if $x \leq y$ in **Z** then $x + z \leq y + z$ for all $z \in \mathbf{Z}$;
(O2)   if $x \leq y$ and $z \geq 0$ in **Z** then $xz \leq yz$.

(Note that $a \geq b$ means $b \leq a$. Also $a < b$, or $b > a$, means that $a \leq b$ and $a \neq b$. The integer a is called *positive* if $a > 0$ and *negative* if $a < 0$.)

III. *The well-ordering principle:* every non-empty set of positive integers has a least member. (A least member of a set S of integers is an integer $m \in S$ such that $m \leq s$ for every $s \in S$.)

The reader may, if he wishes, take the statements I, II, III as axioms for **Z**; they characterize it completely as far as its algebraic structure is concerned. An alternative but lengthy procedure is to start from Peano's axioms for the set **N** of natural numbers and build **Z** from that (see, for example, Halmos: *Naive Set Theory*, (Princeton University Press, 1960), pp. 46–53). We shall simply assume I, II, III without further discussion and deduce some further properties of **Z**. First, some easy consequences of I and II alone; in these assertions a, b, c stand for arbitrary integers, and the universal quantifiers are omitted in accordance with the convention adopted in Chapter 2.

(i) $ab = 0 \Rightarrow a = 0$ or $b = 0$;
(ii) $a < b \Leftrightarrow \neg\, (b \leq a)$;
(iii) $a < b \Rightarrow a + c < b + c$;
(iv) $a < b \Leftrightarrow b - a > 0$;
(v) $a < b$ and $c > 0 \Rightarrow ac < bc$;
(vi) $a^2 \geq 0$;
(vii) $1 > 0$.

The first of these is an alternative form of the cancellation law (M3′). To prove it, we need only assume that $ab = 0$ and $a \neq 0$, and deduce that $b = 0$. This follows from (M3′) directly, since $ab = 0 = a \cdot 0$. (See Chapter 1 for a proof that $a \cdot 0 = 0$. The proof does not use M3.) To prove (ii), observe that if $a < b$ then certainly $a \leq b$. If it were also true that $b \leq a$ then we would have $a = b$ by the anti-symmetry of $\leq$. But $a \neq b$; hence the supposition $b \leq a$ must be false, and we have proved $a < b \Rightarrow \neg\, (b \leq a)$. Conversely, suppose that $\neg\, (b \leq a)$. Then, because $\leq$ is a *linear* order relation, we must have $a \leq b$. Also, $b \neq a$ (since if $b = a$ then $b \leq a$ is true). Thus $a < b$ and we have proved $\neg\, (b \leq a) \Rightarrow a < b$. The first half of the proof of (ii) is a good example of the use of *reductio ad absurdum*: in order to prove that P implies Q it is sufficient to suppose that P is true and Q is false and derive a contradiction or false statement from this supposition.

To prove (iii) it is enough to prove that

$$\neg(a + c < b + c) \Rightarrow \neg(a < b).$$

(The reader should satisfy himself that the statements $P \Rightarrow Q$ and $\neg\, Q \Rightarrow \neg\, P$ mean the same thing.) Using (ii), it is enough to prove that $a + c \geq b + c \Rightarrow a \geq b$, and this follows from (O1) since $a + c \geq b + c \Rightarrow (a + c) + (-c) \geq (b + c) + (-c)$.

We leave the proofs of (iv) and (v) as exercises and turn to (vi). Since the ordering is linear we have, for any given integer a, either $a \geq 0$ or $a \leq 0$. If $a \geq 0$ then $a \cdot a \geq 0$ by (O2). On the other hand, if $a \leq 0$ then, by (O1), $0 = a + (-a) \leq 0 + (-a) = -a$, so $-a \geq 0$ and we have $(-a)(-a) \geq 0$. But $(-a) \cdot (-a) = a \cdot a$ (see the end of Chapter 1 for a proof of this). Hence in both cases we have $a^2 = a \cdot a \geq 0$. Since $1 = 1^2 \geq 0$ and $1 \neq 0$ it follows immediately that $1 > 0$.

There are many similar statements about integers, familiar to the reader, which are easy consequences of I and II. We shall take all such statements for granted now, because to prove them would be tedious and repetitious. For the most part they are straightforward and unlikely to be misunderstood, but there is one common mistake which must receive special mention. It is *not* true that $a \leq b$ implies $ac \leq bc$ for all $a, b, c \in \mathbf{Z}$. According to (O2), this implication is true for $c \geq 0$, but multiplication by negative integers actually *reverses* the ordering. The reader should prove (from I and II), as a compulsory exercise, that

$a \leq b$ and $c \leq 0 \Rightarrow ac \geq bc$

and $a < b$ and $c < 0 \Rightarrow ac > bc$.

He should also record these facts indelibly in his memory.

The consequences of assumption III are much deeper and we shall discuss them more fully. We shall see that this well-ordering principle is closely related to the principle of induction which provides one of the most powerful methods of proof available to mathematicians.

First we note two statements similar to the well-ordering principle and easily derived from it. We say that a subset S of $\mathbf{Z}$ is bounded above if $(\exists b \in \mathbf{Z})(\forall s \in S)(s \leq b)$, and bounded below if $(\exists b' \in \mathbf{Z})(\forall s \in S)(s \geq b')$. The number b satisfying the first of these formulae is called an upper bound of S, and similarly b' is a lower bound of S. We can now state:

(a) every non-empty subset of S of $\mathbf{Z}$ which is bounded below has a least member;
(b) every non-empty subset S of $\mathbf{Z}$ which is bounded above has a largest member.

The proof of (a) consists of adding a suitable fixed integer n to all the members of S so that they all become positive; this is possible because S is bounded below. The resulting set has a least member m, by III, and $m - n$ is the least member of S. The proof of (b) consists of reversing the ordering; for example, let S' be the set of all integers $-s$, where $s \in S$. Then S' is bounded below and has a least member m. The integer $-m$ is then the largest member of S. We leave the details as an exercise.

The statements (a) and (b) must not be confused with similar statements which the student may have met in connection with real numbers. The statements (a) and (b) are false if $\mathbf{Z}$ is replaced by $\mathbf{R}$; for example, the set of positive real numbers has no least member (although it does have a greatest lower bound). Of course, the set of all positive *integers* does have a least member and we can prove:

(c)   1 is the least positive integer.

This statement certainly cannot be deduced from I and II alone because there are other systems of numbers satisfying I and II (e.g. $\mathbf{Q}$, $\mathbf{R}$) in which 1 is not the least positive number. To prove (c), let m be the least positive integer, which exists by III. Then $m \leq 1$ since 1 *is* a positive integer. It is enough, therefore, to suppose that $m < 1$ and obtain a contradiction. Now if $m < 1$ then, since $m > 0$, we have $m \cdot m < 1 \cdot m$, that is, $m^2 < m$. Also $m^2 > 0$ since $m > 0$. Thus m^2 is a smaller positive integer than m, which contradicts the definition of m.

THEOREM 3A. (The Principle of Induction). *Let $P(n)$ be a proposition containing a variable n which stands for an integer. Suppose that $P(1)$ is true and that, for every integer $n \geq 1$, $P(n) \Rightarrow P(n + 1)$. Then $P(n)$ is true for all positive integers n.*

Proof. We argue by contradiction. Suppose that the conclusion is false. Then the set S of all positive integers n for which $P(n)$ is false, is not empty, and therefore it has a least member m. Clearly $m \neq 1$ since $P(1)$ is true. So $m > 1$, since 1 is the *least* positive integer. It follows that $m - 1$ is positive and less than m. By the definition of m it follows that $P(m - 1)$ must be true. Also, since $m - 1 \geq 1$ we have $P(m - 1) \Rightarrow P(m)$. Hence $P(m)$ is true, and we have arrived at a contradiction. It follows that the conclusion of the theorem is true.

There are, of course, many variants of the principle of induction. For example, if $P(0)$ is true and $(P(n) \Rightarrow P(n + 1)$ for all $n \geq 0)$, then $P(n)$ is true for all $n \geq 0$. The essential things are that there should be a starting point for the induction, that is, $P(a)$ must be known to be true for some a, and that there should be no gaps in the chain of implications $P(a) \Rightarrow P(a + 1) \Rightarrow P(a + 2) \Rightarrow \ldots$, i.e., the implication $P(n) \Rightarrow P(n + 1)$ must hold for all $n \geq a$. Under these assumptions $P(n)$ is true for all $n \geq a$ and this is proved by a similar argument.

In practice it is important to write out inductive proofs with some care to avoid incorrect arguments. The procedure is: (i) prove $P(1)$; (ii) suppose that $P(n)$ is true for some $n \geq 1$ (this supposition is referred to as the *induction hypothesis*); (iii) deduce $P(n + 1)$ from this induction hypothesis; (iv) appeal to the principle of induction to conclude that $P(r)$ is true for all integers $r \geq 1$. It is sometimes more convenient to replace (ii) by a stronger induction hypothesis, namely that $P(r)$ is true for all positive integers $r < n$, and then deduce $P(n)$. This method is justified by:

THEOREM 3B. (The Principle of Induction, second form). *Let $P(n)$ be as in Theorem 3A. Let $Q(n)$ be the proposition "$P(r)$ is true for all integers $1 \leq r < n$". Suppose that $P(1)$ is true and that for all $n \geq 1$, $Q(n) \Rightarrow P(n)$. Then $P(n)$ is true for all positive integers n.*

Proof. Let S and m be as in the proof of Theorem 3A. Then $m \neq 1$ and for all integers r satisfying $1 \leq r < m$, $P(r)$ is true. Thus $Q(m)$ is true and hence $P(m)$ is true, a contradiction as before.

Example 3.1. To prove that

$$\sum_{i=1}^{n} i^2 = \frac{1}{6}n(n + 1)(2n + 1)$$

for all positive integers n. First, the proposition is true when $n = 1$

since the left-hand side is

$$\sum_{i=1}^{1} i^2 = 1^2 = 1$$

and the right-hand side is

$$\frac{1}{6} \cdot 1 \cdot 2 \cdot 3 = 1.$$

Assume, as induction hypothesis, that the proposition is true when $n = r$, i.e.,

$$\sum_{i=1}^{r} i^2 = \frac{1}{6} r(r+1)(2r+1).$$

Then

$$\sum_{i=1}^{r+1} i^2 = \frac{1}{6} r(r+1)(2r+1) + (r+1)^2$$

$$= \frac{1}{6}(r+1)[r(2r+1) + 6(r+1)]$$

$$= \frac{1}{6}(r+1)(2r^2 + 7r + 6)$$

$$= \frac{1}{6}(r+1)(r+2)(2r+3).$$

This is the value of

$$\frac{1}{6} n(n+1)(2n+1)$$

when $n = r + 1$, so the proposition is true for $n = r + 1$. The argument is valid for all $r \geq 1$ so, by the principle of induction, the formula is true for all $n \geq 1$.

Example 3.2. To prove that every integer $n \geq 2$ can be written as a product of (one or more) prime numbers. (A prime number can be defined to be an integer ≥ 2 which cannot be written as a product of two smaller positive integers.) This proposition is true for $n = 2$ since 2 is itself prime. We use the method of Theorem 3B. Suppose, as induction hypothesis, that every integer r satisfying $2 \leq r < n$ is a product of primes, and consider n. Either it is itself prime or it is a product $n = n_1 n_2$ where $2 \leq n_1 < n$, $2 \leq n_2 < n$. In the latter case, by induction hypothesis each of n_1, n_2 is a product of primes and hence so is $n = n_1 n_2$. The result now follows by the principle of induction.

Example 3.3. As promised in Chapter 2, we now show that for positive integers m, n the sets $M = \{1, 2, \ldots, m\}$ and $N = \{1, 2, \ldots, n\}$ are not similar unless $m = n$. Suppose that there is a bijection $f : M \to N$; we shall use induction on m to show that $m = n$. If $m = 1$ it is clear that $n = 1$ also. For $m > 1$, we let $f(m) = r$. Then $r \in N$ and the map $g : N \to N$ defined by $g(r) = n$, $g(n) = r$ and $g(x) = x$ for all other $x \in N$, is a bijection (since it is its own inverse). Hence $h = g \circ f$ is a bijection from M to N. Now $h(m) = g(f(m)) = g(r) = n$ and therefore h induces by restriction a bijection from $\{1, 2, \ldots m - 1\}$ to $\{1, 2, \ldots, n - 1\}$. By induction hypothesis, this implies that $m - 1 = n - 1$ and hence that $m = n$. The result now follows by induction.

The method of induction is useful, not only for proving propositions, but also for making iterative definitions. A typical example is the definition of $n!$ for arbitrary positive integers n. It is a function f defined on the set of positive integers by the inductive rules:

$$f(1) = 1; \quad f(n + 1) = (n + 1)f(n) \quad \text{for } n \geqslant 1.$$

The usual loose definition: $n! = 1 \cdot 2 \cdot 3 \cdots n$ is a convenient shorthand for this inductive definition. (The student should be aware that any occurrence of dots $\cdots$ or phrases such as "and so on", "etc." is likely to hide an inductive procedure which may or may not be easy to justify.) The validity of the inductive definition of $n!$ above lies in the fact that there exists a unique function f on the positive integers satisfying the stated conditions. This can be proved from the well-ordering principle by much the same argument as we used in Theorem 3A. It is tricky, however, to formulate and prove a precise statement which takes care of all such inductive definitions, and we shall not attempt this. We refer the interested reader to the theory of recursive functions in, for example, Halmos, *Naive Set Theory*, p. 43.

A good example of an inductive proof which is best carried out by using the well-ordering principle directly instead of Theorem 3A or 3B is the proof of the following basic fact about the process of division in $\mathbf{Z}$.

THEOREM 3C. (The Euclidean property of $\mathbf{Z}$). *Let a, $b \in \mathbf{Z}$, with $b > 0$. Then there exist q, $r \in \mathbf{Z}$ such that $a = bq + r$ and $0 \leqslant r < b$. Furthermore, q and r are unique subject to these properties.*

Proof. Let T be the set of all integers of the form $a - bq$, where q is an arbitrary integer and a, b are the given, fixed, integers. Clearly $T \neq \emptyset$, and we show first that T contains at least one element $t \geqslant 0$. For if $a \geqslant 0$ then $t = a = a - b \cdot 0$ is such an element. On the other hand, if $a < 0$, then since $b \geqslant 1$, we have $ba \leqslant a$; so $t = a - ba$ is such an element. Thus the set $S = \{t \in T; \ t \geqslant 0\}$ is a non-empty set of integers $\geqslant 0$ and so has a least member. Let r be this least member. Then

$r = a - bq$ for some q, and $r \geq 0$. It remains to show that $r < b$. If not, then $r \geq b$ and so $r - b \geq 0$. But $r - b = a - bq - b = a - b(q + 1)$ is in T and hence in S. Furthermore since $b > 0$, $r - b < r$, so we have found a member of S smaller than r. This contradiction proves that $r < b$. Finally, to show the uniqueness of q and r, suppose that we also have $a = bq' + r'$ and $0 \leq r' < b$. We want to show that $q = q'$ and $r = r'$. Now if $q \neq q'$ then either $q > q'$ or $q < q'$ and, by symmetry, we may assume $q > q'$. Since $a = bq + r = bq' + r'$, we have $r' - r = b(q - q') \geq b$ since $b \geq 0$ and $q - q' \geq 1$. This implies that $r' \geq b + r \geq b$, contradicting our assumptions; so we must have $q = q'$. It follows immediately that $r = r'$.

The number r in Theorem 3C is known as the *remainder* of a on division by b, or the *residue of a modulo b*. It is uniquely determined once a and b are given and takes one of the values $0, 1, 2, \ldots, b - 1$. If we now fix a positive integer n and consider division by n, there are exactly n possible residues $0, 1, 2, \ldots, n - 1$ that an integer a can have modulo n. Those integers which have a given residue r modulo n form a subset of **Z** which is called a *residue class modulo n*. There are exactly n such classes (clearly all non-empty) and they form a partition of **Z**. They are in fact the fibres of the function which assigns to each integer its residue modulo n. (See Example 2.23.)

The residue classes modulo n, because they form a partition of **Z**, are the equivalence classes of a suitable equivalence relation on **Z** (Theorem 2D). We now describe this equivalence relation in its nicest form.

Definition. We say that the integer t *divides* the integer s if there exists an integer u such that $s = ut$. (Note, in particular, that every integer divides 0, but 0 divides no integer other than itself.)

To say that t divides s is the same as to say that the residue of s modulo t is 0. The notation used for "t divides s" is $t \mid s$. The notation $t \nmid s$ means "t does not divide s".

We now define, for fixed $n > 0$, a relation $\equiv$ on **Z** called *congruence modulo n*. For integers x, y we write

$$x \equiv y \pmod{n}$$

and say "x is congruent to y modulo n", if n divides $x - y$.

THEOREM 3D. *Congruence modulo n is an equivalence relation and its equivalence classes are the n residue classes modulo n.*

Proof. It is enough to show that $x \equiv y \pmod{n}$ if and only if x and y have the same residue modulo n. Suppose that x and y both have residue r. Then $x = nq + r$ and $y = nq' + r$ for some q, q'. It follows that $x - y = n(q - q')$, so $x \equiv y \pmod{n}$. Conversely, if $x \equiv y \pmod{n}$, then $x = y + nt$ for some $t \in \mathbf{Z}$. If r is the residue of y modulo n, then $y = nq + r$ for some $q \in \mathbf{Z}$ and hence $x = n(q + t) + r$. Since $0 \leq r < n$ we

38   THE INTEGERS

see that r is also the residue of x modulo n (by the uniqueness part of Theorem 3C).

The theory of congruences will be a recurrent theme of this book. For the present we need only the notation and the fact that congruence modulo n is an equivalence relation. The quotient set of $\mathbf{Z}$ associated with this equivalence relation is denoted by $\mathbf{Z}_n$. Its members are the residue classes modulo n and it is therefore a *finite* set with exactly n members. We shall always denote by $\langle x \rangle$ the residue class containing x, writing $\langle x \rangle_n$ if the modulus n is in doubt.

Exercises

1. Prove directly that congruence modulo n is an equivalence relation, without using the Euclidean property of $\mathbf{Z}$.

2. Prove that for every integer x, either $x \geqslant 0$ or $-x \geqslant 0$.

3. For an integer x define
$$|x| = \begin{cases} x & \text{if } x \geqslant 0 \\ -x & \text{if } x < 0. \end{cases}$$
   Prove:
   (i) $|x| \geqslant 0$ for all $x \in \mathbf{Z}$;
   (ii) $|xy| = |x| \, |y|$ for all $x, y \in \mathbf{Z}$;
   (iii) $|x + y| \leqslant |x| + |y|$ for all $x, y \in \mathbf{Z}$.

4. Prove, by induction, that the following statements are true for all positive integers n:
   (i) $\sum_{i=1}^{n} (-1)^i i^2 = \frac{1}{2}(-1)^n n(n+1)$;
   (ii) $\sum_{i=1}^{n} i(i+1) = \frac{1}{3} i(i+1)(i+2)$;
   (iii) $n^2 \leqslant 2^n$;
   (iv) the number of distinct subsets (including the empty subset) of a set of n elements is 2^n,
   (v) $\sum_{i=1}^{n} i! < (n+1)!$

   (Note that $i!$, $(-1)^n$, 2^n and even $\sum_{i=1}^{n}$ are defined inductively.)

5. Let n be a fixed positive integer. Prove that, for every $a \in \mathbf{Z}$, there exists an integer m such that $mn > a$. (This can be deduced easily

from the Euclidean property of **Z** but is really more basic and should preferably be deduced directly from assumptions I, II and III.)

6. Prove that if $n \mid m$ then there is a function $f : \mathbf{Z}_m \to \mathbf{Z}_n$ such that, for every integer x,

$$f(\langle x \rangle_m) = \langle x \rangle_n.$$

CHAPTER 4

Groups

We are now ready to begin our study of abstract algebra. The reader will have noticed that in the laws of "standard algebra" (Example 1.1) the addition laws (A1)–(A4) bear a strong resemblance to the multiplication laws (M1)–(M4). To get from one to the other, one only needs to replace + by ×, 0 by 1, $-x$ by x^{-1}. This suggests that the study of the laws (A1)–(A4) in isolation might be profitable. In fact the commutative law (A4) is irrelevant for many purposes (for example, neither (A4) nor (M4) was used in the specimen proofs given at the end of Chapter 1), so it pays to try and do without it for as long as possible. The remaining three laws (A1)–(A3) have, over the past 150 years, acquired an ever-increasing importance in mathematics. Systems obeying these laws are called "groups" and they occur not only in the context of standard algebra but in almost every corner of mathematics. We shall display a variety of examples as soon as we have given the formal definition.

A *group* consists of a set G, a binary operation $\circ$ on G, a unary operation * on G and a special element $e \in G$, satisfying the following laws:

(G1) $(x \circ y) \circ z = x \circ (y \circ z)$ for all $x, y, z \in G$;

(G2) $e \circ x = x \circ e = x$ for all $x \in G$;

(G3) $x \circ x^* = x^* \circ x = e$ for all $x \in G$.

The element e is called the *neutral element* of G, because of property (G2). The element x^* is called the *inverse of x* with respect to the operation $\circ$, because of property (G3). If a group satisfies the extra law

(G4) $x \circ y = y \circ x$ for all $x, y \in G$,

then it is called a *commutative group,* or an *Abelian group* (after the Norwegian mathematician Abel, one of the founders of group theory).

The symbols $\circ$, *, e in this definition have been chosen to avoid bias in favour of addition or multiplication. Any other symbols would do just as well. In particular we may, when convenient, use additive or multiplicative notation. In additive notation we use + instead of $\circ$ for the binary operation, $-x$ instead of x^*, and 0 instead of e. The laws then become (A1)–(A3) of Example 1.1. In this notation it is usual to abbreviate $x + (-y)$ to $x - y$ so that $-$ becomes a binary operation. A group written in this notation is called an additive group and its neutral element is called its zero element. In multiplicative notation we use × or · instead of $\circ$, and $^{-1}$ instead of *. The neutral element is denoted variously by 1, e, I, etc., according to context. This notation is often simplified by omitting the symbol for the binary operation altogether, writing xy for the "product" of x and y. A group written thus is called a multiplicative group, and its neutral element is often called its identity element.

A group is called finite or infinite according as the set G has finitely many members or not. If it is finite, the number of elements in G is called the *order* of the group. If it is infinite, one says that the group has infinite order.

Example 4.1. The set **Z**, with operations +, − and neutral element 0 is a group. Similarly, **Q**, **R** and **C** are additive groups (with respect to the usual meaning of +, −, 0).

Example 4.2. The set of even integers is an additive group. So is the set of all rational numbers of the form $n/2$, where $n \in$ **Z**. In these examples one must check (as always) that addition and negation actually define operations on the given sets (for example, the sum of two even integers is an even integer) and that 0 belongs to the given sets.

Example 4.3. The set of all non-zero elements of **Q** (respectively, **R**, **C**) is denoted by **Q*** (respectively **R***, **C***). Since products and inverses of non-zero elements are non-zero it is clear that **Q***, **R*** and **C*** are multiplicative groups. So are **Q**^{pos} and **R**^{pos}, the sets of all *positive* rational numbers and real numbers, respectively.

Example 4.4. The above groups are all infinite, but there are also some finite multiplicative groups composed of complex numbers. Let P_n denote the set of all nth roots of 1 in **C**. Since $x^n = y^n = 1 \Rightarrow (xy)^n = 1$, and $x^n = 1 \Rightarrow (x^{-1})^n = 1$, we see that multiplication and inversion are operations on P_n. Also $1 \in P_n$ and the group laws hold in P_n, so P_n is a multiplicative group. Since the number of nth roots of 1 in **C** is exactly n, we see that P_n is a group of order n. For example $P_1 = \{1\}$, $P_2 = \{1, -1\}$ and $P_4 = \{1, -1, i, -i\}$ are multiplicative groups.

Example 4.5. The set T of all complex numbers z of modulus 1 is a multiplicative group. (Note that if $|z| = |w| = 1$ then $|zw| = |z||w| = 1$, so $z, w \in T \Rightarrow zw \in T$, and similarly $z \in T \Rightarrow z^{-1} \in T$.) This group is called the *circle group* because of its geometrical interpretation in the Argand plane. All the groups described so far are Abelian.

Example 4.6. Our next example has nothing to do with numbers and illustrates well the universality of the abstract definition of groups. We start with a quite arbitrary set A, and consider the set $\mathscr{S}(A)$ of all bijections from A to A. Such bijections are called *permutations of A* because, in the case of a finite set whose elements are exhibited in a particular order, each bijection gives a recipe for rearranging the elements in a new order using each element exactly once. Now composition of functions defines a binary operation $\circ$ on $\mathscr{S}(A)$ (by Theorem 2B(iii)). This operation is associative (Theorem 2A) and the identity function ι_A is neutral for it. Furthermore, each bijection $f : A \to A$ has (by Theorem 2C) an inverse $f^{-1} : A \to A$ which is also a bijection and satisfies $f \circ f^{-1} = f^{-1} \circ f = \iota_A$. Thus $\mathscr{S}(A)$ is a group with respect to these operations. It is called the *symmetric group on A*. If A happens to be a finite set with n elements, say $A = \{1, 2, 3, \ldots, n\}$, then the symmetric group on A is denoted by $\mathscr{S}_n$; it is a finite group and its order is $n!$ (the number of different permutations of $\{1, 2, \ldots, n\}$). The group $\mathscr{S}_n$ is not Abelian unless $n = 1$ or 2.

Example 4.7. If in the last example we take A to be the real plane $\mathbf{R}^2$ then certain special types of permutations are of geometrical interest. For example, the rigid motions (or Euclidean transformations) are those bijections $\rho : \mathbf{R}^2 \to \mathbf{R}^2$ that preserve distances, i.e. such that for any two points P, Q, the distance from $\rho(P)$ to $\rho(Q)$ is equal to the distance PQ. The set of all such rigid motions is a (non-Abelian) group with respect to composition and inversion of functions; it is called the two-dimensional *Euclidean group*.

Example 4.8. Those rigid motions of $\mathbf{R}^2$ that leave the origin fixed are called orthogonal transformations. They comprise all rotations about the origin and all reflections about lines through the origin, and they form a group with respect to the same operations of composition and inversion of functions. This group is called the *orthogonal group* O_2. The rotations alone also form a group O_2^+ (the special orthogonal group); but the reflections alone do not form a group, because the composite of two reflections is a rotation, not a reflection. The special orthogonal group O_2^+ is Abelian but the full orthogonal group is not.

Example 4.9. If one takes any geometrical figure in the plane, say a collection of points and lines, then a rigid motion of the plane is called a *symmetry* of the figure if it maps the figure into itself. The individual points and lines need not be fixed, but a symmetry permutes the points of the figure and permutes its lines. The symmetries of any figure form a group (with respect to composition of functions). One may also form other groups by selecting those symmetries of the figure that belong to the orthogonal group or the special orthogonal group. For example, a square centred at the origin has 8 symmetries, of which 4 are rotations and 4 are reflections. They form a non-Abelian group. The 4 rotational symmetries also form a group, but this group is Abelian.

Example 4.10. The set of all $n \times n$ real matrices is a group with respect to $+$, $-$; the zero element of this group is the matrix whose entries are all 0. Multiplication of matrices is an associative binary operation on this set, and the identity matrix I_n (with diagonal entries 1 and all other entries 0) is neutral for multiplication, but since matrices do not usually have inverses, this is not a group. However, if we restrict attention to the set of all *invertible* (non-singular) $n \times n$ matrices we obtain a multiplicative group known as the *general linear group* $GL_n(\mathbf{R})$. It is not Abelian unless $n = 1$. Note that the set of invertible matrices is *not* an additive group because the sum of invertible matrices is not necessarily invertible.

The examples above show that groups are very common in many mathematical contexts. Any theorems we prove about groups will be applicable to all these examples, and many more. The surprising thing is that so much comes out of such common-place assumptions. In proving theorems about arbitrary groups we shall normally use multiplicative notation and denote the neutral element by e. The corresponding theorems for additive groups may be obtained (and proved) by a simple substitution of symbols; the meaning of a theorem does not depend on the particular notation used.

THEOREM 4A. *Let G be a group. Then*

(i) *the neutral element is unique*;

(ii) *for each $x \in G$ there is a unique $y \in G$ such that $xy = yx = e$*;

(iii) *for $x, y \in G$, $xy = e \Rightarrow x = y^{-1}$ and $y = x^{-1}$*;

(iv) *$(xy)^{-1} = y^{-1}x^{-1}$ for all $x, y \in G$*;

(v) *$(x^{-1})^{-1} = x$ for all $x \in G$*;

(vi) *(cancellation laws) for $a, x, y \in G$*,

$$ax = ay \Rightarrow x = y$$

and $\quad xa = ya \Rightarrow x = y$;

(vii) *if $x_1, x_2, \ldots, x_n \in G$ then the product $x_1 x_2 \ldots x_n$ is independent of the position of brackets*;

(viii) *if $x_1, x_2, \ldots, x_n \in G$ and G is Abelian then the product $x_1 x_2 \ldots x_n$ is independent of the order of the factors as well as the bracketing.*

Proof. (i) If e, e' are two neutral elements then $e = ee' = e'$.

(ii) If y, y' are two elements with the given property then $y = ye = y(xy') = (yx)y' = ey' = y'$. But for each x the element x^{-1} has the property $xx^{-1} = x^{-1}x = e$, so there is exactly one such element.

(iii) Suppose that $xy = e$. Then $x^{-1}(xy) = x^{-1}e = x^{-1}$. But $x^{-1}(xy) = (x^{-1}x)y = ey = y$. So $y = x^{-1}$. Similarly, $x = xe = x(yy^{-1}) = (xy)y^{-1} = ey^{-1} = y^{-1}$.

(iv) $(xy)(y^{-1}x^{-1}) = ((xy)y^{-1})x^{-1} = (x(yy^{-1}))x^{-1} = (xe)x^{-1} = xx^{-1} = e$. But, by (iii), this implies that each of the elements xy and $y^{-1}x^{-1}$ is inverse to the other. In particular $y^{-1}x^{-1} = (xy)^{-1}$.

(v) $xx^{-1} = e$ by (G3). But, by (iii), this implies that $x = (x^{-1})^{-1}$.

(vi) If $ax = ay$ then $a^{-1}(ax) = a^{-1}(ay)$. Hence $(a^{-1}a)x = (a^{-1}a)y$, that is, $x = y$. Similarly, $xa = ya \Rightarrow x = y$.

(vii) This is proved by induction on n. For $n = 1, 2$ it is obvious. For $n = 3$ it is the associative law (G1). The argument for $n = 4$ has been partly carried out in (iv) above; it consists of numerous applications of the associative law. We now give the argument for general n. Let P be any bracketed product of $x_1, x_2, \ldots, x_n$, in that order. Then $P = Q_1 S_1$, where Q_1, S_1 are products of length less than n, and the two factors are uniquely determined by the bracketing. If S_1 has length 2 or more, then it is itself a product $S_1 = R_2 S_2$, so $P = Q_1(R_2 S_2) = (Q_1 R_2)S_2 = Q_2 S_2$, say. Clearly S_2 is shorter than S_1. If S_2 has length 2 or more we may repeat this procedure until the second factor has length 1. (There is a subsidiary inductive argument hidden here.) Thus we have $P = Qx_n$, where Q is a bracketed product of $x_1, x_2, \ldots, x_{n-1}$. Similarly, if P' is any other bracketed product of $x_1, x_2, \ldots, x_n$, then $P' = Q'x_n$, where Q' is a product of $x_1, x_2, \ldots, x_{n-1}$. By induction hypothesis, $Q = Q'$, and so $P = Qx_n = Q'x_n = P'$. The result now follows for all n by induction.

(viii) The argument here is similar, using the commutative law many times to rearrange the product without altering its value. By (vii) we need not write any brackets, and this simplifies the notation. The most obvious procedure is to prove that any product of $x_1, x_2, \ldots, x_n$ is equal to the particular product $x_1 x_2 \ldots x_n$. This can be done as follows. First move x_1 to the front by successive interchanges with its predecessors, each interchange being justified by the commutative law. One can then deal similarly in turn with $x_2, x_3, \ldots, x_n$, moving each to its proper place. More briefly, we may use induction on n. For $n = 1, 2$ the result is obvious. If we move x_1 to the front as above, then we may appeal to our induction hypothesis to justify the rearrangement of $x_2, x_3, \ldots, x_n$ in correct order, and the result follows.

Some immediate consequences of these elementary results should be noted. Firstly, by (vii), we can always, if we wish, omit brackets when writing products in a group; the notation $x_1 x_2 \ldots x_n$ (or $x_1 + x_2 + \ldots + x_n$ in an additive group) is unambiguous. Similarly, by (viii), we need not specify the order of the factors in a product *provided that the group is Abelian.* The notation

$$\prod_{i=1}^{n} x_i \quad (\text{or } \sum_{i=1}^{n} x_i \text{ in an additive group})$$

is therefore permissible if the group is known to be Abelian, but not otherwise.

Secondly, parts (i) and (ii) of the theorem show that when we wish to assert that something is a group we need only specify the set and the binary operation on it. If it is a group, the neutral element, and the inverse of each element, are uniquely determined once the binary operation is known. This fact justifies an alternative definition of "group" which is commonly used: a group is a set G with a binary operation $\circ$ defined on it such that (i) $\circ$ is associative, (ii) there is an element $e \in G$ satisfying $e \circ x = x \circ e = x$ for all $x \in G$ and (iii) for each $x \in G$ there is an element $y \in G$ such that $x \circ y = y \circ x = e$. We shall in future use such descriptions as "G is a group with respect to $+$" or simply "$(G, +)$ is a group" without fear of ambiguity.

Thirdly, part (iv) of the theorem implies, by an easy induction that for any product $x_1 x_2 \ldots x_n$ (which we may write without brackets)

$$(x_1 x_2 \ldots x_n)^{-1} = x_n^{-1} x_{n-1}^{-1} \ldots x_1^{-1}.$$

If G is any (multiplicative) group we can define the *powers* of an element $x \in G$ inductively as follows: $x^0 = e$; $x^1 = x$; $x^{n+1} = x^n x$ for $n \geq 1$. We can also define negative powers of x by the rule $x^{-n} = (x^n)^{-1}$ for $n \geq 1$. (Observe that for $n = 1$ the two possible meanings of x^{-1} agree.)

THEOREM 4B. *Let G be any group. Then*

(i) $\quad x^m x^n = x^{m+n}$ *for all* $x \in G$, $m, n \in \mathbb{Z}$;

(ii) $\quad (x^m)^n = x^{mn}$ *for all* $x \in G$, $m, n \in \mathbb{Z}$;

(iii) $\quad$ *if G is Abelian then* $(xy)^n = x^n y^n$ *for all* $x, y \in G$, $n \in \mathbb{Z}$.

Proof. If n is a positive integer then x^n is, by definition, one of the possible products of n factors, all equal to x. It follows that, for positive integers m and n, the statements (i), (ii) and (iii) are special cases of Theorem 4A, (vii) and (viii). (This argument uses intuitively the connection between positive integers and counting procedures. The only way to avoid this appeal to intuition is to base the theory of the natural numbers firmly on Peano's axioms. This would greatly increase the complexity of the proof without adding to the understanding.) If either $m = 0$ or $n = 0$ the statements are trivially true. For other values of m and n we appeal to the definition $x^{-n} = (x^n)^{-1}$ to reduce each statement to the positive case. For example, suppose that $m < 0, n > 0$ and $m + n > 0$ in (i). Put $p = -m > 0$. Then, by the positive case, $x^{m+n} x^p = x^{m+n+p} = x^m$. But $x^n = x^{-p} = (x^p)^{-1}$, by definition, so

$x^m x^n = x^{m+n} x^p x^n = x^{m+n}$. The other cases of (i) are similarly proved. To prove (ii) when $m < 0, n > 0$, put $q = -m > 0$, and first note that $(y^r)^{-1} = (y^{-1})^r$, for any $y \in G$ and $r > 0$, by the formula for the inverse of a product. Hence $(x^m)^n = (x^{-q})^n = ((x^q)^{-1})^n = ((x^q)^n)^{-1} = (x^{qn})^{-1} = x^{-qn} = x^{mn}$. We leave the other cases as an exercise for the reader.

COROLLARY. *In any group the powers of a single element commute with each other.*

Proof. $x^m x^n = x^{m+n} = x^{n+m} = x^n x^m$ because addition of integers is commutative.

When a group G is written additively the "powers" of an element x are written nx instead of x^n (e.g. $3x = x + x + x$ instead of $x^3 = x \cdot x \cdot x$). We shall call them "integer multiples of x" or "additive powers of x". Note that the definition $x^0 = e$ becomes $0x = 0$, in which 0 is used first to denote the integer 0 and then to denote the zero element of G. The exponential laws of Theorem 4B become $mx + nx = (m+n)x$, $n(mx) = (nm)x$ and, *for Abelian groups,* $n(x+y) = nx + ny$. The reader should write out for himself the additive form of Theorem 4A and remember it. We shall use these two theorems very frequently and usually without specific back-reference. They are the rules on which all calculations in groups are based. We should perhaps mention that additive notation is rarely (and never in this book) used for any group which is not Abelian.

Example 4.11. In the group $GL_2(\mathbf{R})$ (see Example 4.10), the element

$$X = \begin{pmatrix} 1 & 1 \\ 0 & 1 \end{pmatrix}$$

has inverse

$$X^{-1} = \begin{pmatrix} 1 & -1 \\ 0 & 1 \end{pmatrix}.$$

The powers of X are

$$X^n = \begin{pmatrix} 1 & n \\ 0 & 1 \end{pmatrix}$$

for all $n \in \mathbf{Z}$. This is proved by induction for $n > 0$ and by inversion for $n < 0$.

Example 4.12. In the group $\mathscr{S}_n$ of permutations of $\{1, 2, \ldots, n\}$, we shall adopt the following notation: the permutation which sends 1 to r_1, 2 to r_2, and generally i to r_i will be denoted by

$$r = \begin{pmatrix} 1 & 2 & \cdots & n \\ r_1 & r_2 & \cdots & r_n \end{pmatrix}.$$

This notation helps the computation of products; we give some examples from $\mathscr{S}_3$ and leave the reader to work out his own technique:

$$\begin{pmatrix} 1 & 2 & 3 \\ 2 & 1 & 3 \end{pmatrix} \begin{pmatrix} 1 & 2 & 3 \\ 1 & 3 & 2 \end{pmatrix} = \begin{pmatrix} 1 & 2 & 3 \\ 2 & 3 & 1 \end{pmatrix},$$

$$\begin{pmatrix} 1 & 2 & 3 \\ 1 & 3 & 2 \end{pmatrix} \begin{pmatrix} 1 & 2 & 3 \\ 2 & 1 & 3 \end{pmatrix} = \begin{pmatrix} 1 & 2 & 3 \\ 3 & 1 & 2 \end{pmatrix}.$$

Remember that multiplication in $\mathscr{S}_n$ is composition of functions, and that $f \circ g$ is the function obtained by first applying g, then f. The examples show that $\mathscr{S}_3$ is not an Abelian group.

There are six elements in $\mathscr{S}_3$, which we denote by

$$e = \begin{pmatrix} 1 & 2 & 3 \\ 1 & 2 & 3 \end{pmatrix}, \quad a = \begin{pmatrix} 1 & 2 & 3 \\ 1 & 3 & 2 \end{pmatrix}, \quad b = \begin{pmatrix} 1 & 2 & 3 \\ 3 & 2 & 1 \end{pmatrix},$$

$$c = \begin{pmatrix} 1 & 2 & 3 \\ 2 & 1 & 3 \end{pmatrix}, \quad p = \begin{pmatrix} 1 & 2 & 3 \\ 2 & 3 & 1 \end{pmatrix}, \quad q = \begin{pmatrix} 1 & 2 & 3 \\ 3 & 1 & 2 \end{pmatrix}.$$

The three *transpositions* a, b, c, each interchange two numbers and are clearly their own inverses, i.e., $a^2 = b^2 = c^2 = e$. Thus $a^n = a$ if n is odd and $a^n = e$ if n is even. The *cyclic permutations* p and q are inverse to each other, and $p^2 = q$, $p^3 = e$. Hence p^n is always one of the three elements p, q, e. We record here, for future reference, a complete multiplication table for $\mathscr{S}_3$. A typical entry is the product xy where x appears at the left of its row and y appears at the head of its column.

	e	a	b	c	p	q
e	e	a	b	c	p	q
a	a	e	p	q	b	c
b	b	q	e	p	c	a
c	c	p	q	e	a	b
p	p	c	a	b	q	e
q	q	b	c	a	e	p

The fact that each element of $\mathscr{S}_3$ appears exactly once in each row and each column of this table is not an accident; it is true for any group and is a consequence of the fact that for fixed elements s, t in a group the equation $sx = t$ (or $xs = t$) has a unique solution.

A group G is called *cyclic* if all its elements are powers of one particular element g, which is then called a *generator* of G. For example the group P_n of nth roots of 1 in $\mathbf{C}$ is cyclic since all its elements are powers of $\zeta = e^{2\pi i/n}$. The additive group $\mathbf{Z}$ is also cyclic since every integer is an additive power of 1. The group $\mathscr{S}_3$, on the other hand, is not cyclic because no element has more than 3 distinct powers (see Example 4.11). Any cyclic group is Abelian because all the powers of its generator commute with each other (Theorem 4B, Corollary). Note that the generator of a cyclic group is not unique; for example -1 is a generator of the additive group $\mathbf{Z}$, and $e^{4\pi i/5}$ is a generator of P_5. We shall have more to say about this later.

A *subgroup* of a group G is a subset H of G such that

(i) $e \in H$;
(ii) $x, y \in H \Rightarrow xy \in H$;
(iii) $x \in H \Rightarrow x^{-1} \in H$.

The reason for imposing these conditions is that they ensure that H is itself a group. They say that H has the necessary special element e and that the multiplication and inversion in G induce a binary and a unary operation on H when their domains are restricted. Since the laws (G1), (G2), (G3) hold for all elements of G they are valid automatically in H and so H is a group with respect to the induced operations. Conversely, it is easy to see that if a subset H is a group with respect to the restricted multiplication then it must satisfy (i) and (iii) above, as well as (ii), so our definition has in fact caught all the multiplicative groups contained in G.

Example 4.13. In any group G the subsets $\{e\}$ and G are subgroups.

Example 4.14. If x is any fixed element of a group G, then the set X whose members are all the powers of x, is a subgroup, called *the cyclic subgroup generated by x*. The reasons are: (i) $e = x^0$; (ii) $x^m x^n = x^{m+n}$; (iii) $(x^n)^{-1} = x^{-n}$.

Example 4.15. It is not difficult to find all the subgroups of $\mathscr{S}_3$. We refer to Example 4.12 for the multiplication table of $\mathscr{S}_3$, and we can immediately write down all the cyclic subgroups of $\mathscr{S}_3$. They are $\{e\}$, $\{e, a\}$, $\{e, b\}$, $\{e, c\}$ and $\{e, p, q\}$. Let H be any subgroup not in this list. If $H \supset \{e, p, q\}$ then H contains at least one of a, b, c. By symmetry we can assume that $a \in H$, and then H contains $ap = b$ and $aq = c$, so $H = \mathscr{S}_3$. On the other hand, if $H \not\supset \{e, p, q\}$ then $p \notin H$, $q \notin H$, so H must contain at least two of a, b, c. But this implies that $p \in H$ since $p = ab = bc = ca$. This contradiction shows that $\mathscr{S}_3$ is the only subgroup other than the cyclic subgroups already listed.

In an *additive* group A the cyclic subgroup generated by an element x consists of all the *additive* powers nx of x. In particular in the additive group $\mathbf{Z}$ of all integers, the cyclic subgroup generated by the integer d is the set $\mathbf{Z}d = \{nd; n \in \mathbf{Z}\}$. Here the set $\mathbf{Z}$ is playing a dual role. On the one hand it has replaced A as the group under consideration. On the other hand we are using integers $n \in \mathbf{Z}$ to indicate the nth additive power of an element d, as we can do in any additive group. It so happens that for this very special group the notation nd already has a meaning as the *product* of two integers. Fortunately the two meanings coincide, since for positive n the product nd is $d + d + \ldots + d$ (n times). Both interpretations appear in the proof of our next theorem which is fundamental for the development of arithmetic in Chapter 5.

THEOREM 4C. *Every subgroup H of the additive group $\mathbf{Z}$ is cyclic, i.e. is of the form $H = \mathbf{Z}d$ for some $d \in \mathbf{Z}$. The generator d can be chosen so that $d \geqslant 0$ and is then uniquely determined by H.*

Proof. If $H = \{0\}$ we may take $d = 0$ and the statement is obviously true. So suppose that $H \neq \{0\}$. Then H contains some $h \neq 0$, and since, as a subgroup, it must also contain $-h$ it must contain at least one positive member. By the well-ordering principle, H contains a *least* positive member d (see p. 31). Clearly all additive powers of d are in H since H is a subgroup, so $\mathbf{Z}d \subset H$. Now suppose that $n \in H$. By the Euclidean property of $\mathbf{Z}$ (Theorem 3C), $n = qd + r$, where $q, r \in \mathbf{Z}$ and $0 \leqslant r < d$. Since H is a subgroup we have $qd \in H$, and hence $r = n - qd \in H$. But d is the least positive member of H, and $0 \leqslant r < d$, so r is not positive, and must be 0. Thus $n = qd \in \mathbf{Z}d$ and it follows that $H = \mathbf{Z}d$. We have shown that d can be chosen positive or 0. The uniqueness of d is obvious when $H = \{0\}$. If $H \neq \{0\}$, suppose that $c > 0$ is a generator of H. Then all positive members of H are of the form nc with $n > 0$. But this implies $n \geqslant 1$ and hence $nc \geqslant c$. Thus c is the smallest positive member of H and is therefore uniquely determined by H.

COROLLARY. *If G is a cyclic group then every subgroup of G is cyclic.*

Proof. Let G be generated by g and let K be any subgroup. Define $H = \{n \in \mathbf{Z}; g^n \in K\}$. Then H is an additive subgroup of $\mathbf{Z}$ (check the three conditions!). Hence, by the theorem, $H = \mathbf{Z}d$ for some d. Now $k = g^d \in K$, since $d \in H$, and it is clear that k generates K. For if $g^n \in K$ then $n \in H$ (by definition), so $n = rd$ for some integer r, and $g^n = g^{rd} = k^r$.

In any group, an element x is said to have *finite order* if $x^n = e$ for some $n > 0$; the least such n is then called the *order* of x. If no such n

exists we say that x has *infinite order*. The relationship between the orders of elements and the orders of groups (as previously defined) is given in our next theorem.

THEOREM 4D. *Let G be a group and let $x \in G$. Then the order of x is equal to the order of the cyclic subgroup X generated by x. If x has infinite order, then all its powers are distinct: $x^r = x^s \Leftrightarrow r = s$. If x has finite order, then* (i) $x^k = e \Leftrightarrow n|k$, *and* (ii) $x^r = x^s \Leftrightarrow r \equiv s \pmod{n}$.

Proof. First suppose that x has infinite order, that is, $x^k \neq e$ for $k > 0$. If $x^r = x^s$, and if $r \geq s$ (as we may assume without loss of generality), then $x^{r-s} = x^r(x^s)^{-1} = e$, and $r - s \geq 0$. It follows that $r - s = 0$, i.e., $r = s$. Thus all powers of x are distinct and X is an infinite group.

Next, suppose that x has finite order n, that is, $n > 0$, $x^n = e$, and $x^k \neq e$ for $0 < k < n$. For any integer k, we may write $k = qn + t$, where q and t are integers and $0 \leq t < n$. Since $x^k = x^{qn+t} = (x^n)^q x^t = x^t$ we see immediately that $x^k = e \Leftrightarrow x^t = e \Leftrightarrow t = 0 \Leftrightarrow n|k$. Hence also $x^r = x^s \Leftrightarrow x^{r-s} = e \Leftrightarrow n|r-s \Leftrightarrow r \equiv s \pmod{n}$. Again, the fact that $x^k = x^t$ implies that every power of x is equal to one of the elements $e, x, x^2, \ldots, x^{n-1}$, and these are distinct because no two of the integers $0, 1, \ldots, n - 1$ are congruent modulo n. Thus $X = \{e, x, x^2, \ldots, x^{n-1}\}$ has exactly n elements in this case.

This theorem gives a clear indication of what types of cyclic groups are possible. The infinite ones all look alike, with their elements labelled uniquely by integers. The finite ones of given order n also look alike, with elements $e, x, x^2, \ldots, x^{n-1}$, where x is any generator. We now make precise the phrase "look alike". Two groups A and B are said to be *isomorphic* (written $A \cong B$) if there exists a bijection $f : A \to B$ which preserves the group multiplication, that is, $f(xy) = f(x)f(y)$ for all $x, y \in A$. Such a map f is called an *isomorphism* of groups. It follows automatically from this condition that f also preserves the neutral element and preserves inverses, that is, $f(e) = e$ (where e is used somewhat improperly to denote the neutral elements of both A and B) and $f(x^{-1}) = f(x)^{-1}$ for all $x \in A$. To see this, put $b = f(e)$. Then $b^2 = f(e)f(e) = f(e^2) = f(e) = b$, from which it follows that b is the neutral element of B. Again, put $f(x^{-1}) = c$. Then $cf(x) = f(x^{-1})f(x) = f(x^{-1}x) = f(e) = e$ in B, and this implies $c = f(x)^{-1}$.

It is clear from the definition that the composite of two isomorphisms is an isomorphism. Furthermore, the inverse of an isomorphism (recall that an inverse exists for any bijection) is also an isomorphism. For if $f : A \to B$ is an isomorphism with inverse $g : B \to A$, and if $u, v \in B$, then $u = f(x)$, $v = f(y)$ where $x = g(u)$ and $y = g(v)$. Hence $uv = f(x)f(y) = f(xy)$, and so $g(uv) = g(f(xy)) = xy = g(u)g(v)$. These facts (together with the trivial fact that the identity map on a group is an isomorphism) show that the relation $A \cong B$ between groups is an equivalence relation, that is, for all groups A, B, C, (i) $A \cong A$, (ii) $A \cong B \Rightarrow B \cong A$, and (iii) $A \cong B$ and $B \cong C \Rightarrow A \cong C$. Thus we may classify groups so that two groups are in the same class if and only if they are isomorphic. (See Theorem 2D.) We call these classes "isomorphism classes" or "isomorphism types".

Since multiplication, inversion and neutral element, which are the building blocks for all equations in groups, are preserved by isomorphisms, it follows that if $f : A \to B$ is an isomorphism, then every true equation in A has a counterpart in B. For example, if $a^5 = e$ in A and if $b = f(a)$, then $b^5 = e$ in B, and conversely since f^{-1} is an isomorphism. Thus two isomorphic groups have exactly the same group-theoretical properties and are indistinguishable from this point of view. In looking at the isomorphism type of a group one is ignoring the nature of the individual elements and concentrating on the "algebraic structure", that is, the pattern formed by the algebraic relations between the elements.

If one (or both) of the groups A, B is written in another notation, the definition must be modified accordingly. Thus an isomorphism from an additive group to a multiplicative group is a bijection f satisfying $f(x + y) = f(x)f(y)$. It will then automatically satisfy $f(0) = e$ and $f(-x) = f(x)^{-1}$. Its inverse g will satisfy $g(uv) = g(u) + g(v)$, $g(e) = 0$ and $g(u^{-1}) = -g(u)$.

Example 4.16. The additive group **Z** is isomorphic with the additive group of all even integers by means of the isomorphism $n \mapsto 2n$.

Example 4.17. The set G of all matrices

$$\begin{pmatrix} a & b \\ -b & a \end{pmatrix},$$

where a and b are real numbers, not both zero, is a group with respect to matrix multiplication. This group G is isomorphic with the multiplicative group **C*** of non-zero complex numbers. The complex number corresponding to the matrix

$$\begin{pmatrix} a & b \\ -b & a \end{pmatrix}$$

is $a + bi$, and the reader should verify that this correspondence is in fact an isomorphism.

Example 4.18. The additive group **R** is isomorphic with the multiplicative group **R**^{pos} of *positive* real numbers by means of the exponential map $x \mapsto e^x$. That this is a bijection depends on results in elementary analysis. So does the well-known formula $e^{x+y} = e^x e^y$ which states that the map is an isomorphism. It follows that $e^0 = 1$

and $e^{-x} = (e^x)^{-1}$; also that the inverse function, called log, satisfies $\log(1) = 0$, $\log(uv) = \log u + \log v$ and $\log(u^{-1}) = -\log u$. This isomorphism is, of course, the basis for the use of logarithms to multiply positive numbers. The fact that the two groups are isomorphic tells us that, in spite of appearances, multiplication of positive real numbers is essentially the same algebraic operation as addition of real numbers. All we need in order to change the one to the other is a method of applying the isomorphism, e.g., a set of logarithm tables, or a slide-rule.

THEOREM 4E. *Any two cyclic groups of the same order are isomorphic.*

Proof. Let X, Y be two cyclic groups, generated by x, y, respectively. If both are of infinite order then, by Theorem 4D, all powers of x are distinct and so are all powers of y. It follows that the map $f : X \to Y$ defined by $f(x^r) = y^r$ is a bijection. It is an isomorphism because $f(x^r x^s) = f(x^{r+s}) = y^{r+s} = y^r y^s = f(x^r)f(x^s)$. Now suppose that both X and Y are of finite order n. Then by Theorem 4D, $X = \{e, x, x^2, \ldots x^{n-1}\}$ and $Y = \{e, y, y^2, \ldots, y^{n-1}\}$, so we can define a bijection $f : X \to Y$ by the rule $f(x^r) = y^r$ for $0 \leq r < n$. The rule for multiplication in X is as follows. If $0 \leq r < n$, $0 \leq s < n$, then $x^r \cdot x^s = x^{r+s} = x^t$ where

$$\begin{cases} t = r + s & \text{if } r + s < n, \\ t = r + s - n & \text{if } r + s \geq n. \end{cases}$$

The rule is the same for Y, so it is clear that f preserves multiplication and is therefore an isomorphism.

COROLLARY. (i) *Every infinite cyclic group is isomorphic with the additive group* **Z**.

(ii) *Every finite cyclic group of order n is isomorphic with the group* P_n *of nth roots of* 1 *in* **C**.

We end this chapter by proving a famous theorem of Lagrange which asserts that in a finite group of order n, the order of every subgroup must be a divisor of n. It is proved by partitioning the group into subsets all of the same size, and the partition arises as follows. Let H be a subgroup of a group G (not necessarily finite). For $x, y \in G$ write $x \sim y$ to mean that $xy^{-1} \in H$ (or in additive notation $x - y \in H$). Then $\sim$ is an equivalence relation on G, the three conditions for an equivalence relation corresponding in a remarkable way to the three conditions for a subgroup. The connection between the two concepts is obviously very close:

(i) $\sim$ is reflexive because $xx^{-1} = e \in H$, i.e. $x \sim x$.

(ii) $\sim$ is symmetric because if $x \sim y$ then $xy^{-1} \in H$, so $(xy^{-1})^{-1} \in H$. But $(xy^{-1})^{-1} = (y^{-1})^{-1}x^{-1} = yx^{-1}$, so $y \sim x$.

(iii) $\sim$ is transitive because if $x \sim y$ and $y \sim z$ then $xy^{-1} \in H$ and $yz^{-1} \in H$, so $xz^{-1} = (xy^{-1})(yz^{-1}) \in H$, i.e. $x \sim z$.

The equivalence classes arising from this relation are called *right cosets* of H in G. The right coset $\langle x \rangle$ containing the element x consists of all elements $y \in G$ such that $yx^{-1} \in H$, that is, $y = hx$ for some $h \in H$. Thus $\langle x \rangle = Hx$, where (in the obvious notation) Hx means $\{hx; h \in H\}$. The right coset containing e is $He = H$.

A similar argument leads to the *left cosets* of H in G. We start with the relation $x^{-1}y \in H$ and find that it is an equivalence relation whose classes are the sets $xH = \{xh; h \in H\}$. Of course, if G is an Abelian group then left and right cosets coincide. In additive notation we write $H + x$ and $x + H$ for the right and left cosets.

Example 4.19. Let G be the additive group $\mathbf{C}$ and let H be the subgroup $\mathbf{R}$. Then the right (and left) cosets of H are (in the Argand diagram) the lines parallel to the real axis. More generally if H is the subgroup consisting of all real multiples of a fixed complex number z, which is a line through 0, then the cosets of H are the lines parallel to it. They partition the plane.

Example 4.20. Let G be the multiplicative group $\mathbf{C}^*$ and let H be the circle group T (see Example 4.5), which is a subgroup of $\mathbf{C}^*$. Then the right (and left) cosets of T are the circles with centre 0. For, if z_0 is a fixed complex number then the coset Tz_0 consists of all complex numbers $z = tz_0$ where $|t| = 1$, and these numbers are precisely those satisfying $|z| = |z_0|$.

Example 4.21. Let $G = \mathbf{C}^*$ and let H be the multiplicative group $\mathbf{R}^{pos}$ of positive real numbers. Then the cosets of H in G are the half-lines radiating from 0 in the Argand diagram. (The one containing a given complex number w consists of all positive real multiples of w.)

Example 4.22. If G is the additive group $\mathbf{R}$ and H is the subgroup $\mathbf{Z}$, then a typical coset $\mathbf{Z} + x$ consists of the points $n + x$ ($n = 0, \pm 1, \pm 2, \ldots$) which are distributed along the real line at unit distance apart.

Example 4.23. If G is the additive group $\mathbf{Z}$ and H is the subgroup $n\mathbf{Z}$ then a coset $n\mathbf{Z} + r$, where $r \in \mathbf{Z}$, consists of all integers $a = nq + r$ for $q \in \mathbf{Z}$. In other words, the coset $n\mathbf{Z} + r$ is the residue class modulo n containing r. Indeed, in this case the equivalence relation defined by the subgroup $n\mathbf{Z}$ is just congruence modulo n, because $a - b \in n\mathbf{Z} \Leftrightarrow n \mid a - b$.

Example 4.24. To show that left and right cosets do not always coincide we take $G = \mathscr{S}_3$, the symmetric group on three symbols, and $H = \{e, a\}$, where a is the transposition

$$\begin{pmatrix} 1 & 2 & 3 \\ 1 & 3 & 2 \end{pmatrix}.$$

Referring to the notation and the multiplication table in Example 4.12, we see that the distinct right cosets of H are H, $Hp = \{p, b\}$ and $Hq = \{q, c\}$, whereas its left cosets are H, $pH = \{p, c\}$ and $qH = \{q, b\}$.

THEOREM 4F. *Let H be a subgroup of the group G. Then*

(i) *the right cosets Hx of H form a partition of G;*

(ii) $Hx = Hy \Leftrightarrow x \in Hy \Leftrightarrow y \in Hx$;

(iii) *there is a bijection between any two right cosets of H. A similar result holds for left cosets.*

Proof. We have already shown that Hx is the equivalence class containing x for the equivalence relation defined by $x \sim y \Leftrightarrow xy^{-1} \in H$. Statement (i) follows immediately. It also follows that $Hx = Hy \Leftrightarrow x \sim y \Leftrightarrow xy^{-1} \in H \Leftrightarrow x \in Hy$, and symmetry gives $Hx = Hy \Leftrightarrow y \in Hx$. Finally, to prove (iii), it is enough to show that there is a bijection from H to Hx, for any $x \in G$. (The statement will then follow from Theorems 2B and 2C.) But it is easy to see that the map $f : H \to Hx$ defined by $h \mapsto hx$ (for fixed x) is a bijection. It is an injection because $hx = h'x \Rightarrow h = h'$ by the cancellation law for groups; it is a surjection because, by definition, every element of Hx is of the form hx for some $h \in H$.

COROLLARY 1. *(Lagrange's Theorem) If the group G is finite, of order n, then the order of every subgroup is a divisor of n.*

Proof. If H is a subgroup of order m, then each right coset of H contains exactly m elements, by part (iii) of the theorem. Since the right cosets form a partition of G we have $n = rm$, where r is the number of cosets.

COROLLARY 2. *If the group G is finite, of order n, then $x^n = e$ for every $x \in G$.*

Proof. Let $x \in G$ and let X be the cyclic subgroup generated by x. If the order of X is m, then $x^m = e$, by Theorem 4D. But $n = rm$ for some integer r, by Langrange's Theorem. Hence $x^n = x^{rm} = (x^m)^r = e^r = e$.

COROLLARY 3. *Any group of prime order is cyclic.*

Proof. Let G be a group of prime order p. Since $p \geqslant 2$, there is an element $x \neq e$ in G, and it generates a cyclic subgroup X with at least two elements. The order m of X divides p, by Langrange's Theorem, and since $m \neq 1$ we must have $m = p$. Hence $X = G$ and G is cyclic.

Warning. The converse of Lagrange's Theorem is false; that is, if a group G has finite order n and if m is a divisor of n, there may not be any subgroups of order m. In certain special circumstances one can however prove the existence of a subgroup of order m; for example, if G is cyclic, this is true for all m dividing n (see Theorem 7F). More generally, if G is Abelian, one can show that subgroups of order m exist for every m dividing n. A famous theorem of Sylow's also asserts that if m is a *prime-power* dividing n, then a subgroup of order m exists. The last two results will be found in most standard texts on group theory but are outside the scope of this introductory book.

Exercises

1. Which of the following are groups? Give reasons.

    (i)   The set of all odd integers with respect to addition.

    (ii) The set of all rational numbers of the form $m/2^n$ with respect to (a) addition and (b) multiplication.

    (iii) The set of all real numbers except -1 with respect to the operation $*$ defined by $a * b = a + b + ab$.

    (iv) The set of all 2×2 matrices of the form
    $$\begin{pmatrix} a & 0 \\ b & c \end{pmatrix},$$
    where $b \in \mathbf{Q}$, $a, c \in \mathbf{R}$, $a \neq 0$, $c \neq 0$, the operation being multiplication of matrices.

2. Which of the following groups are cyclic? Give reasons.

    (i)   The additive group $\mathbf{Q}$ of rational numbers.

    (ii) The circle group $T = \{ z \in \mathbf{C}; |z| = 1 \}$ (see Example 4.5).

    (iii) The symmetric group $\mathscr{S}_3$ (see Example 4.12).

    (iv) The group of rotational and reflectional symmetries of a rectangle.

    (v)   The general linear group $\mathrm{GL}_n(\mathbf{R})$ for arbitrary $n \geqslant 1$ (see Example 4.10).

3. Four functions $\alpha, \beta, \gamma, \delta : \mathbf{R}^* \to \mathbf{R}^*$ are defined by $\alpha(x) = x$, $\beta(x) = -x$, $\gamma(x) = x^{-1}$, $\delta(x) = -x^{-1}$. Prove that they form a group under composition of functions, and that this group is isomorphic with the group of rotational and reflectional symmetries of a rectangle (not a square).

4. Prove that if $x^2 = e$ for every element x of a group G, then G is Abelian.

5. Prove that every group of order less than or equal to 5 is Abelian.

6. The following is part of the multiplication table of a group of order 6. Fill in the gaps.

	p	q	r	s	t	u
p	r	·	·	t	·	·
q	·	t	·	·	r	·
r	p	·	·	·	·	·
s	·	·	·	r	·	·
t	·	·	·	p	·	·
u	·	s	·	·	·	r

7. Separate the following groups into isomorphism classes and justify your classification.

  (i) The circle group $T = \{z \in \mathbf{C}; |z| = 1\}$.

  (ii) The multiplicative group of sixth roots of 1 in $\mathbf{C}$.

  (iii) The symmetric group $\mathscr{S}_3$.

  (iv) The additive group $\mathbf{Z}$.

  (v) The group of rotational symmetries of a regular hexagon.

  (vi) The group of rotational and reflectional symmetries of an equilateral triangle.

  (vii) The group of functions $f_n : \mathbf{Z} \to \mathbf{Z}$ defined by $f_n(x) = x + n$, with respect to composition of functions. (There is one function f_n in the group for each integer n.)

  (viii) The multiplicative group of all real numbers of the form 2^n ($n \in \mathbf{Z}$).

  (ix) The additive group $\mathbf{R}$.

  (x) The multiplicative group of all matrices of the form

$$\begin{pmatrix} \cos\theta & \sin\theta \\ -\sin\theta & \cos\theta \end{pmatrix}$$

where $\theta \in \mathbf{R}$.

(xi) The additive group $\mathbf{Q}$.

(xii) The multiplicative group of all matrices of the form

$$\begin{pmatrix} 1 & x \\ 0 & 1 \end{pmatrix}$$

where $x \in \mathbf{R}$.

(xiii) The special orthogonal group O_2^+. (See Example 4.8.)

8. Prove that the multiplicative group $\mathbf{Q}^{pos}$ is not isomorphic with the additive group $\mathbf{Q}$.

9. Find all the subgroups of a cyclic group of order 12.

10. If A and B are subgroups of a group G, show that $A \cap B$ is also a subgroup. Is $A \cup B$ necessarily a subgroup? (Give reasons).

11. Let G be a group and let $g \in G$ be a fixed element. Prove that $H = \{x \in G; x^{-1}gx = g\}$ is a subgroup of G.

12. Prove that in any Abelian group G, the elements of finite order form a subgroup. Identify this subgroup when (i) $G = O_2^+$, (ii) $G = \mathbf{Q}$ (additive), (iii) $G = \mathbf{Q}^*$ (multiplicative), (iv) $G = \mathbf{C}^*$ (multiplicative).

13. Prove that if G is a group whose only subgroups are G and $\{e\}$ then G is cyclic of prime order. (Do not assume that G is finite, and remember that the converse of Lagrange's theorem is not true.)

14. (Harder). Prove that if p and q are distinct prime numbers, then any group of order pq is cyclic.

15. (Harder). Find all the subgroups of $\mathscr{S}_4$.

16. (Harder). Let G be a finite group of order n and let A, B be subgroups of orders a, b, respectively. Prove that the order of $C = A \cap B$ (see Exercise 10) is at least ab/n. (Hint: show that each right coset of C is contained in a right coset of B and that those which are contained in A all lie in different cosets of B.)

17. (Harder). Prove that the number of ways of bracketing a product of n symbols is

$$\frac{1}{n}\binom{2n-2}{n-1}.$$

CHAPTER 5

Factorization in Z

Most people learn at an early stage of their training in arithmetic that if a natural number is factorized repeatedly it can be expressed as a product of prime numbers, and the prime factors which appear are independent of the way in which the factorization is performed. This important principle is easily verified for small numbers and it is natural at this early stage to assume that it remains true for large numbers. The principle is used later for computing the greatest common divisor and the least common multiple of two numbers — this is done by comparing and combining the prime factorizations of the numbers. Least common multiples are used, of course, to simplify calculations with fractions.

There are two criticisms of this procedure which a serious student of mathematics should ponder. Firstly, it is by no means obvious that a very large number whose prime factors are not known (and perhaps not even knowable with present-day computers) is uniquely expressible as a product of primes. Such a sweeping statement demands a proof, using only simple statements of a more intuitive nature. Secondly, the factorization of a number into its prime constituents is computationally a very complicated procedure, and one ought to ask whether a simpler and shorter algorithm exists for computing greatest common divisors. In fact both proof and algorithm were known to Euclid over 2000 years ago and it is a great pity that they are not more widely taught today.

Euclid's proof of unique factorization into primes, which we shall describe shortly, uses the existence and properties of greatest common divisors; it is essential, therefore, that these should be established first *without reference to prime factors*. The procedure outlined in the first paragraph above, although it may be the best way to teach the subject initially, is logically unsound since it puts the cart before the horse. It also fails to reveal the interesting and important fact that the greatest common divisor of a and b can always be written in the form $ma + nb$, where m and n are integers. We prove this as a by product of the proof of the existence of greatest common divisors, and we base the proof on group theory, because it is the group structure which best explains the reason for the appearance of integers of the form $ma + nb$.

Our initial assumptions for $\mathbf{Z}$ have been described in Chapter 3. We recall that, for $x, y \in \mathbf{Z}$, $x \mid y$ means "x divides y", i.e. $(\exists z \in \mathbf{Z})(y = xz)$, and we note some easy consequences:

(i) $x \mid 0$ for all $x \in \mathbf{Z}$;

(ii) if $0 \mid a$ then $a = 0$;

(iii) if $a \mid 1$ then $a = 1$ or $a = -1$;

(iv) if $x \mid y$ and $y \mid x$ then $y = x$ or $y = -x$.

Of these, (i) and (ii) follow immediately from the definition. To prove (iii), suppose that $a \mid 1$, i.e. $ab = 1$ for some integer b. Clearly $a \neq 0$ and $b \neq 0$, so either a and b are both positive or they are both negative. If they are both positive then $a \geqslant 1$ and $b \geqslant 1$, so $1 = ab \geqslant a$ and we deduce that $a = 1$ since 1 is the least positive integer. Similarly, if they are both negative, then $a \leqslant -1, b \leqslant -1$, so $1 = ab \geqslant a(-1) = -a > 0$, and we deduce that $a = -1$. Statement (iv) follows easily because if $x = my$ and $y = nx$ ($m, n \in \mathbf{Z}$) then $x = mnx$. It follows that either $x = 0$ or $mn = 1$, by the cancellation law for $\mathbf{Z}$. In the first case, $y = nx = 0$, so $y = x$. In the second case, by (iii), $n = \pm 1$ and so $y = nx = \pm x$.

Another way of saying $a \mid 1$ is to say that a is *invertible* in $\mathbf{Z}$, that is, it has an inverse a^{-1} in $\mathbf{Z}$ such that $aa^{-1} = 1 = a^{-1}a$. Statement (iii) above then asserts that the only invertible elements of $\mathbf{Z}$ are 1 and -1. These invertible elements are also called the *units* of $\mathbf{Z}$.

Given two integers a, b, we say that an integer d is *a greatest common divisor* of a and b if

(i) $d \mid a$ and $d \mid b$,

(ii) $(\forall c \in \mathbf{Z})((c \mid a$ and $c \mid b) \Rightarrow c \mid d)$, and

(iii) $d \geqslant 0$.

Conditions (i) and (ii) say that d is a common divisor which is divisible by *all* common divisors. Condition (iii) is added for convenience, in order to make d unique. Note that the act of *defining* a new term does not in any way force either its existence or its uniqueness; these have to be proved before the definition acquires its real usefulness.

THEOREM 5A. *Any two integers a and b have a unique greatest common divisor d. Further, d can be written in the form d = ra + sb for suitable r, s $\in$ **Z**.*

Proof. The uniqueness is easy: if d and d' are two greatest common divisors of a and b then, by definition, $d \mid d'$ and $d' \mid d$. Hence $d' = \pm d$ and, since $d \geqslant 0$ and $d' \geqslant 0$, it follows that $d = d'$. Thus a and b have *at most one* greatest common divisor. To show that they have *at least one*, we consider the set $H = \{ma + nb; m, n \in \mathbf{Z}\}$. (Here a and b are fixed, m and n vary). This H is a subset of $\mathbf{Z}$ and we now verify that it is an additive *subgroup* of $\mathbf{Z}$. There are three conditions to check, and they

all follow easily from the definition of H:
- (i) $0 \in H$, since $0 = 0a + 0b$;
- (ii)   if $x \in H$ then $x = ma + nb$, so $-x = (-m)a + (-n)b \in H$;
- (iii)   if $x, x' \in H$ then $x = ma + nb$, $x' = m'a + n'b$, so $x + x' = (m + m')a + (n + n')b \in H$.

But we know what the additive subgroups of $\mathbf{Z}$ are: by Theorem 4C, H is cyclic, that is, $H = \mathbf{Z}d = \{zd; z \in \mathbf{Z}\}$ for some $d \geqslant 0$, and we claim that this integer d is actually a greatest common divisor of a and b. In the first place, since $H = \mathbf{Z}d$, every member of H is divisible by d. In particular, since $a = 1a + 0b \in H$ and $b = 0a + 1b \in H$, we have $d \mid a$ and $d \mid b$. In the second place, since $d \in H$ we have $d = ra + sb$ for some $r, s \in \mathbf{Z}$, and hence, if $c \mid a$ and $c \mid b$ then $c \mid d$ (because, if $a = xc$ and $b = yc$ then $d = ra + sb = (rx + sy)c$). Since $d \geqslant 0$ it satisfies all the conditions for a greatest common divisor. Incidentally we have also proved the last assertion of the theorem.

COROLLARY. *Let $a, b, c \in \mathbf{Z}$. Then the equation $ax + by = c$ has an integral solution [i.e., $(\exists x, y \in \mathbf{Z})(ax + by = c)$] if and only if the greatest common divisor of a and b divides c.*

Proof. Let d be the greatest common divisor of a and b. Then $H = \mathbf{Z}d$ is precisely the set of all integers of the form $ax + by$ for $x, y \in \mathbf{Z}$, as was shown in the proof of the theorem. So $ax + by = c$ has a solution $\Leftrightarrow c \in H \Leftrightarrow d \mid c$. Alternatively, we may use the theorem itself rather than its proof: (i) if $ax + by = c$ for integers x, y, then, since $d \mid a$ and $d \mid b$, we have $d \mid c$; (ii) by the theorem $d = ar + bs$ for some $r, s \in \mathbf{Z}$, so if $d \mid c$ then $c = dt = a(rt) + b(st)$ and we have an integral solution $x = rt$, $y = st$.

Remarks. (1) The subgroup $H = \{ma + nb; m, n \in \mathbf{Z}\}$ should be compared with the cyclic subgroups $A = \{ma; m \in \mathbf{Z}\}$ and $B = \{nb; n \in \mathbf{Z}\}$. It contains them both and is in fact the smallest subgroup of $\mathbf{Z}$ which contains both a and b.

(2) d is not *uniquely* expressible as $ma + nb$.

(3) The proof of the theorem shows that the greatest common divisor of a and b is actually the *smallest* non-negative number of the form $ma + nb$.

(4) The name "greatest common divisor" can now be justified (if a and b are not both 0) because d, being positive, and being divisible by all common divisors, is indeed the largest of them.

(5) If $a = 0$ and b is positive then $d = b$. If $a = 0$ and b is negative, then $d = -b$. The "greatest common divisor" of 0 and 0 is 0, but in this case the name is inappropriate since all integers are common divisors.

(6) The least common multiple m of a and b can be similarly defined by the conditions:

(i) $a \mid m, b \mid m$;
(ii) $(a \mid m'$ and $b \mid m') \Rightarrow m \mid m'$;
(iii) $m \geq 0$.

One can deduce the existence of m from Theorem 5A by showing that if a and b are not both zero and if $a = da'$, $b = db'$, where d is their greatest common divisor, then the integer $ab/d = a'b'd$ has properties (i) and (ii). The uniqueness of m is easy. Alternatively one can imitate the proof of Theorem 5A, replacing H by the subgroup $K = A \cap B$, where A and B are as in Remark (1).

Notation. The greatest common divisor of a and b is denoted by $d = (a, b)$. The reader will be aware that this notation is already overworked, but since all three meanings have become standard in different contexts it is difficult to justify changing any of them. They rarely clash.

Definition. Two integers a and b are *coprime* (or *relatively prime*) if $(a, b) = 1$. This is the same as saying that their only common divisors are the units 1 and -1.

We now collect for future reference some of the main properties of greatest common divisors. The proofs do not use prime factorizations but rely entirely on the definitions and on the equation $d = ma + nb$.

THEOREM 5B. (i) $(ac, bc) = (a, b)c$ for all $a, b, c \in \mathbf{Z}, (c \geq 0)$.
(ii) If $d = (a, b) \neq 0$, then $a = da'$, $b = db'$ where a' and b' are coprime.
(iii) a and b are coprime $\Leftrightarrow (\exists m, n \in \mathbf{Z})(ma + nb = 1)$.
(iv) If $a \mid bc$ and $(a, b) = 1$, then $a \mid c$.
(v) If $x \mid z$ and $y \mid z$ and $(x, y) = 1$, then $xy \mid z$.
(vi) If $a = bq + r$, then $(a, b) = (b, r)$.

Proof. (i) Let $d = (a, b)$. Then $d \mid a$ and $d \mid b$, so $dc \mid ac$ and $dc \mid bc$. Also, d can be written in the form $d = ma + nb$, so $dc = m(ac) + n(bc)$ is divisible by any common divisor of ac and bc. Since $dc \geq 0$, this shows that $dc = (ac, bc)$.

(ii) If $d = (a, b) \neq 0$ then, since $d \mid a$ and $d \mid b$, we can write $a = da'$, $b = db'$, where a' and b' are uniquely determined. By (i), we have $d = (a, b) = (da', db') = d(a', b')$, and since $d \neq 0$, this implies $(a', b') = 1$.

(iii) This is a special case of the Corollary to Theorem 5A. The equation $ma + nb = 1$ has integral solutions m, n if and only if $(a, b) \mid 1$, i.e., a and b are coprime.

(iv) Let $(a, b) = 1$. Then $\exists m, n \in \mathbf{Z}$ such that $ma + nb = 1$. Hence $c = mac + nbc$. Clearly $a \mid mac$, and if $a \mid bc$, we also have $a \mid nbc$, whence $a \mid (mac + nbc) = c$.

(v) Let $z = ax$ and $z = by$. If $(x, y) = 1$ then $1 = mx + ny$ for suitable $m, n \in \mathbf{Z}$. Hence $z = zmx + zny = bymx + axny = (bm + an)xy$.

(vi) Any common divisor of a and b divides $r = a - bq$ and so is a common divisor of b and r. Similarly, any common divisor of b and r is a common divisor of a and b.

Definition. An integer p is *prime* if (i) every divisor of p is of the form u or up where u is a unit (i.e. $u = \pm 1$), (ii) p is not itself a unit, and (iii) $p \geqslant 0$. This definition may seem unfamiliar, but is easily seen to be equivalent to the more usual ones. The reason for phrasing the definition in this particular way will appear when we discuss factorisation of polynomials in Chapter 10.

THEOREM 5C. (i) *If p, q are prime and $p \mid q$ then $p = q$.*

(ii) *If a is an integer, p is prime and $p \nmid a$ then $(p, a) = 1$.*

(iii) *If $m > 1$ is not prime then $\exists a, b \in \mathbf{Z}$ such that $m = ab$, $1 < a < m$ and $1 < b < m$.*

(iv) *If p is prime and $p \mid a_1 a_2 \ldots a_r$, then $p \mid a_i$ for at least one i in the range $1 \leqslant i \leqslant r$.*

Proof. (i) Since p is a divisor of the prime number q, we must have $p = u$ or $p = uq$ for a unit u. Since p is prime, $p = u$ is ruled out, and since p and q are both positive we cannot have $p = -q$. Hence $p = q$.

(ii) Let $d = (p, a)$. Then $d \mid p$ and $d \mid a$. Since $p \nmid a$ we cannot have $d = p$ or $d = -p$. Hence d is a unit and, being positive, it must be 1.

(iii) If $m > 0$ is *not* prime then, by definition, either m is a unit or m has a divisor which is not ± 1 or $\pm m$. If $m > 1$, the first alternative is ruled out, so $m = ab$ where $a \neq \pm 1, a \neq \pm m$. Since also $m = (-a)(-b)$, we may choose $a > 0$ and therefore $b > 0$. Thus $a \geqslant 1, b \geqslant 1$ and consequently $m = ab \geqslant b$ and $m = ab \geqslant a$. We now have $m = ab$, $1 \leqslant a \leqslant m$, $1 \leqslant b \leqslant m$. But $a \neq 1, a \neq m$, so $1 < a < m$, and it follows that $1 < b < m$.

(iv) This is the most important property of prime numbers. We use induction on r. If $r = 1$, there is nothing to prove. If $r > 1$ then $p \mid aa_r$, where $a = a_1 a_2 \ldots a_{r-1}$. If $p \mid a$, we can apply the induction hypothesis to deduce that $p \mid a_i$ for some i in the range $1 \leqslant i \leqslant r - 1$. On the other hand, if $p \nmid a$, then $(p, a) = 1$, by (ii). Since $p \mid aa_r$ we can deduce that $p \mid a_r$ in this case (Theorem 5B, (iv)). This completes the inductive step from $r - 1$ to r, and the result follows.

We now have all the material necessary for the proof of the main theorem of this chapter, sometimes called the "fundamental theorem of arithmetic", but more informatively called the "unique factorization theorem for **Z**".

THEOREM 5D. (i) *Every non-zero integer can be factorized in the form $n = up_1 p_2 \ldots p_r$, where $u = \pm 1$, each p_i is prime, and $r \geqslant 0$.*

(ii) *If* $n = vq_1 q_2 \ldots q_s$ *is another factorization of* n *with* $v = \pm 1$, q_j *prime, and* $s \geq 0$, *then* $u = v$, $r = s$, *and the prime numbers* $q_1, q_2, \ldots, q_s$ *are a permutation of* $p_1, p_2, \ldots, p_s$.

Proof. (i) Since $n = u \cdot |n|$, where $u = \pm 1$, we may assume that n is positive. (See Chapter 3, Exercise 3 for the definition and properties of $|n|$.) If $n = 1$ we may take $u = 1$ and $r = 0$, and the result is true in this case. So it is enough to take $n > 1$ and show that n is a product of (one or more) prime numbers. Suppose that this is not the case for some $n > 1$. Then by the well-ordering principle there is a *smallest* integer $m > 1$ which cannot be written as a product of prime numbers. Since, in particular, m is not prime we may apply Theorem 5C (iii) to obtain a factorization $m = ab$ with $1 < a < m$, $1 < b < m$. But m is the smallest offending integer, so both a and b *can* be written as products of (one or more) prime numbers, whence so can $m = ab$. This contradiction establishes the result.

(ii) Suppose that $n = up_1 p_2 \ldots p_r = vq_1 q_2 \ldots q_s$. If $n > 0$ then $u = v = 1$, and if $n < 0$ then $u = v = -1$, because all the prime factors are positive. Hence it is enough to consider the case $n > 0$. If $n = 1$, then clearly $r = s = 0$ since no prime number divides 1. We may therefore assume that $n > 1$ and $n = p_1 p_2 \ldots p_r = q_1 q_2 \ldots q_s$ (with $r \geq 1$, $s \geq 1$). Since multiplication of integers is commutative as well as associative we may assume that the prime factors have been permuted so as to appear in ascending order: $p_1 \leq p_2 \leq \ldots \leq p_r$ and $q_1 \leq q_2 \leq \ldots \leq q_s$. Under these circumstances we shall prove that $r = s$ and $p_i = q_i$ for $i = 1, 2, \ldots, r$. For suppose this fails to be true for some integer $n > 1$; then there is a *smallest* integer $n > 1$ which has two different prime factorizations $n = p_1 p_2 \ldots p_r = q_1 q_2 \ldots q_s$ with $p_1 \leq p_2 \leq \ldots \leq p_r$ and $q_1 \leq q_2 \leq \ldots \leq q_s$. The smallest prime number occurring in these two factorizations is either p_1 or q_1, and we may assume, without loss of generality, that it is p_1. Then $p_1 | n = q_1 q_2 \ldots q_s$, so by Theorem 5C (iv) we have $p_1 | q_i$ for some i, and since p_1 and q_i are both prime this implies that $p_1 = q_i$ (see Theorem 5C(i)). But p_1 is the smallest of the prime factors, so we must have $p_1 = q_1$. The rest is now easy. We must have at least one more factor in each product (because the factorizations are different) and by the cancellation law we obtain $p_2 p_3 \ldots p_r = q_2 q_3 \ldots q_s = n'$ say with $1 < n' < n$. We have therefore found a smaller integer with two different factorizations, which contradicts our choice of n, and completes the proof.

We return now to the problem of computing greatest common divisors. It is easy, on the basis of the unique factorization theorem, to justify the method in which one obtains prime factorizations of the numbers and collects the highest powers of primes dividing both of them. However, this is certainly not the easiest method, except for small numbers whose prime factors can be written down immediately. Euclid's algorithm is much more economical and repays close study.

Suppose we wish to find (a, b) for two given integers a, b. Since $(a, 0) = 0$, $(a, a) = a$, $(a, b) = (b, a)$ and $(a, b) = (-a, b) = (a, -b) = (-a, -b)$, we need only deal with the case $a > b > 0$. By Theorem 3C, there exist integers q, r such that $a = bq + r$ and $b > r \geqslant 0$; to find them one simply divides a by b and notes the quotient and remainder. Now Theorem 5B (vi) tells us that $(a, b) = (b, r)$, so we have reduced the computation to finding (b, r) for a smaller pair of (non-negative) integers. Repetition of this procedure will eventually give the answer.

Example 5.1. $(1320, 714) = (714, 606) = (606, 108) = (108, 66) = (66, 42) = (42, 24) = (24, 18) = (18, 6) = 6$.

To describe the algorithm more precisely we introduce notation as follows. Let $a_0 = a$, $a_1 = b$ and let q_1, a_2 be the uniquely determined integers such that $a_0 = a_1 q_1 + a_2$ and $0 \leqslant a_2 < a_1$. We then define q_n, a_n inductively as follows: if $a_n > 0$ then q_n and a_{n+1} are the (unique) integers such that $a_{n-1} = a_n q_n + a_{n+1}$ and $0 \leqslant a_{n+1} < a_n$; if $a_n = 0$ the algorithm terminates. Since $a_1 > a_2 > a_3 > \ldots \geqslant 0$, we cannot have $a_n > 0$ for all n, so the algorithm is always finite (with at most b steps), and we obtain a set of equations

$$a_0 = a_1 q_1 + a_2,$$
$$a_1 = a_2 q_2 + a_3,$$
$$\vdots$$
$$a_{r-3} = a_{r-2} q_{r-2} + a_{r-1},$$
$$a_{r-2} = a_{r-1} q_{r-1} + a_r,$$
$$a_{r-1} = a_r q_r,$$

where a_{r+1} is the first of the a's whose value is 0. In this notation, the greatest common divisor of a and b is the last non-zero remainder a_r, because $d = (a, b) = (a_0, a_1) = (a_1, a_2) = (a_2, a_3) = \ldots = (a_r, a_{r+1}) = (a_r, 0) = a_r$. The virtue of writing out the equations in this way is that they can now be used to express d in the form $am + bn$. The penultimate equation gives $d = a_r = a_{r-2} - a_{r-1} q_{r-1}$. If in this we substitute $a_{r-1} = a_{r-3} - a_{r-2} q_{r-2}$, obtained from the next equation up, we find d expressed in the form $d = a_{r-3} s + a_{r-2} t$. Proceeding in this way until we reach the first equation, we eventually obtain an equation of the form $d = a_0 m + a_1 n = am + bn$.

Example 5.2. Let $a = 975, b = 616$. Then the computation is as follows:

$$975 = 616.1 + 359,$$
$$616 = 359.1 + 257,$$

$359 = 257.1 + 102,$
$257 = 102.2 + 53,$
$102 = 53.1 + 49,$
$53 = 49.1 + 4,$
$49 = 4.12 + 1.$

Thus $(a, b) = 1$ and, following the procedure described above, we obtain

$1 = 49 + 4(-12)$
$= 53(-12) + 49.13$
$= 102.13 + 53(-25)$
$= 257(-25) + 102.63$
$= 359.63 + 257.(-88)$
$= 616(-88) + 359.151$
$= 975.151 + 616(-239).$

This would be difficult to find by trial and error! (There may, of course, be a smaller solution).

Euclid's algorithm can be speeded up by modifying Theorem 3C a little. If, given a and b, one looks for the multiple of b whose difference from a is smallest *in absolute value*, one obtains an equation $a = bq + r$ with $|r| \leq \frac{1}{2}b$. Since $(a, b) = (b, r) = (b, |r|)$, the reduction is at least as good as previously and the algorithm must terminate after at most n steps, where $2^n > b$. The modified algorithm can still be used to solve $d = am + bn$.

Example 5.3. Let $a = 1320$, $b = 714$, as in Example 5.1. Then

$1320 = 714.2 - 108,$
$714 = 108.7 - 42,$
$108 = 42.3 - 18,$
$42 = 18.2 + 6,$
$18 = 6.3.$

Hence the greatest common divisor of a and b is 6, and we have

$6 = 42 + 18(-2)$
$= 108.2 + 42(-5)$
$= 714.5 + 108(-33)$
$= 1320.33 + 714(-61).$

It is interesting to compare this with a quite different use of Euclid's algorithm, concerned with real numbers and their approximation by rational numbers. (This is a digression and can be omitted by students who want to get back to the main topic.) It seems probable that Euclid's algorithm first arose in a geometrical context. Early Greek geometers assumed that any two line segments were *commensurable*, that is, were both exact multiples of some smaller line segment, which may be thought of as a "common divisor" of them. To find such a common measure for two segments a and b of which b is the shorter, lay off b as many times as possible along a. If its fits exactly, then b is a common measure. If not, the piece r_1 left over will be shorter than b, and clearly, any common measure of r_1 and b will also measure a exactly. So repeat the process with b and r_1. If r_1 measures b exactly it is a common measure of a and b. If not the piece r_2 left over is shorter than r_1. Thus one obtains shorter and shorter segments $r_1, r_2, \ldots$, and the mistake made by the early geometers was in thinking that such a process must terminate after a finite number of steps.

In modern notation the lengths of the segments are real numbers a and b and, putting $a_0 = a$, $a_1 = b$, we may define a_n and q_n inductively by the rule: if $a_n \neq 0$ then q_n is the largest *integer* such that $a_n q_n \leqslant a_{n-1}$, and $a_{n+1} = a_{n-1} - a_n q_n$. Thus we have the same sequence of equations

$$a_0 = a_1 q_1 + a_2,$$
$$a_1 = a_2 q_2 + a_3,$$
$$\vdots$$

as before, with the difference that $a_n \in \mathbf{R}$, $q_n \in \mathbf{Z}$, $0 \leqslant a_{n+1} < a_n$ and the sequence may go on indefinitely without a_n ever becoming 0. In fact the algorithm terminates after a finite number of steps if and only if a and b are commensurable, that is, if and only if a/b is a rational number. (Exercise: prove this!). Surprisingly, even when the algorithm fails to terminate, it is not useless; the fact that it is trying, quite efficiently, to find a common measure for a and b means in practice that it supplies us with a series of rational numbers which approximate a/b more and more closely. If one makes an estimate of the efficiency of the algorithm one can prove the following theorem about closeness of approximation of real numbers by rational numbers. *Given any real number m, and given any $\epsilon > 0$, there exist integers p, q such that $q > 0$ and*

$$\left| x - \frac{p}{q} \right| < \frac{\epsilon}{q}.$$

To see this, we put $a = x$ and $b = 1$ and carry out Euclid's algorithm in the more efficient version in which $a_{n-1} = a_n q_n + a_{n+1}$ and

$$|a_{n+1}| \leq \frac{1}{2}|a_n|$$

(one simply takes the integer multiple of a_n which is *closest* to a_{n-1}). Then clearly

$$|a_n| \leq \frac{1}{2^{n-1}}$$

and we may therefore choose $n > 1$ so that $|a_n| < \epsilon$. Now, precisely as in the integral case, a_n can be expressed in the form $a_n = a_0 r + a_1 s$, where r, s are *integers*. Hence $a_n = xr + s$, and so $|xr + s| < \epsilon$. We can assume that $r \neq 0$ (for if $r = 0$, then a_n is an integer and so must be zero; in this case x is rational and the theorem is obviously true). Dividing by $|r|$, we have

$$\left| x + \frac{s}{r} \right| < \frac{\epsilon}{|r|}$$

and we may take $q = |r|$ and $p = \pm s$ to establish the theorem.

Note. This result can also be proved by a group-theoretic argument similar to the one used for Theorem 5A. (See Exercise 14 below.)

Exercises

1. Find the greatest common divisor of 180 and 252, and express it in the form $180x + 252y$, where $x, y \in \mathbf{Z}$.

2. Find an integral solution of the equation $966x + 686y = 70$.

3. Prove that if $x, y, u, v \in \mathbf{Z}$ and $xy \mid uv$, then $(x, u)y \mid u(y, v)$. (Try to do this without using the unique factorization theorem.)

4. Let a, b, c be integers such that $ab = c^2$. Prove that if $(a, b) = 1$ then both a and b are squares of integers.

5. Prove that $((a, b), c)) = (a, (b, c))$ for all integers a, b, c.

6. Is it true or false that $(\forall a, b, c \in \mathbf{Z})((ab, c) = (a, c)(b, c))$?

7. Prove (without using the unique factorization theorem) that if $a \mid bc$, then there exist integers x and y such that $x \mid b$, $y \mid c$ and $a = xy$.

8. Let $m, n_1, n_2, \ldots, n_r$ be integers such that $(m, n_i) = 1$ for $i = 1, 2, \ldots, r$. Prove that $(m, n) = 1$, where $n = n_1 n_2 \ldots n_r$. (Hint: if $(a, b) \neq 1$ then a and b have a common prime divisor.)

9. Prove that if $q \in \mathbf{Q}$ and $q^n \in \mathbf{Z}$ for some integer $n > 0$, then $q \in \mathbf{Z}$. Deduce that if p is a prime number then $p^{1/n}$ is irrational for all $n \geq 2$.

10. Let $a = p_1^{\alpha_1} p_2^{\alpha_2} \ldots p_n^{\alpha_n}$, $b = p_1^{\beta_1} p_2^{\beta_2} \ldots p_n^{\beta_n}$, where $p_1, \ldots, p_n$ are distinct prime numbers, $n \geqslant 1$, and $\alpha_i \geqslant 0$, $\beta_i \geqslant 0$ are integers. Prove that $a \mid b$ if and only if $\alpha_i \leqslant \beta_i$ for $i = 1, 2, \ldots, n$. Find a formula for the number of distinct positive divisors of $a = p_1^{\alpha_1} p_2^{\alpha_2} \ldots p_n^{\alpha_n}$.

11. Let n be a square-free positive integer (i.e. n is not divisible by the square of any prime number). Suppose that, for all primes p, $p \mid n \Leftrightarrow (p-1) \mid n$. Prove that $n = 1806$.

12. (i) Is every additive subgroup of $\mathbf{Q}$ cyclic? (ii) Is every additive subgroup of $\mathbf{Q}$ of the form $H = \mathbf{Q}d = \{qd; q \in \mathbf{Q}\}$ for some $d \in \mathbf{Q}$?

13. Prove that $\mathbf{Z}$ contains infinitely many distinct prime numbers. (Hint: suppose that $p_1, p_2, \ldots, p_n$ are all the distinct prime numbers and obtain a contradiction by looking at the prime factorization of $1 + p_1 p_2 \ldots p_n$.)

14. (Harder) Let G be an additive subgroup of $\mathbf{R}$. Prove that either (i) G is cyclic, or (ii) G contains arbitrarily small non-zero elements. (Hint: either G contains a smallest positive member or it does not.)

If x is a real number, show that $G = \{xm + n; m, n \in \mathbf{Z}\}$ is an additive subgroup of $\mathbf{R}$. Show that G is cyclic if and only if x is rational. Hence prove that for any irrational x and for any $\epsilon > 0$ there exist integers p, q such that $q > 0$ and

$$\left| x - \frac{p}{q} \right| < \frac{\epsilon}{q}.$$

15. (Harder) Prove that the probability that two randomly chosen positive integers are coprime is $6/\pi^2 = 0{:}61\ldots$ (Notes: The probability is to be defined by first taking the probability for pairs of integers not exceeding N and then letting $N \to \infty$. A rigorous argument is difficult, but a fairly convincing argument can be obtained by considering the probability that two positive integers should both be divisible by a given prime number p. You will need to know that

$$\sum_{n=1}^{\infty} \frac{1}{n^2} = \frac{\pi^2}{6}.)$$

CHAPTER 6

New Groups From Old

So far we have used groups only in a descriptive role; we have simply introduced a new language for talking about familiar situations, our chief gain being a unification of ideas. However, the crystallization of the abstract idea of a group opens up a new possibility. Instead of just looking at the mathematical objects around us in the hope of finding more groups on which to try the new methods, we can actually manufacture new groups that we have not seen before. In this chapter we describe two constructions (product groups and quotient groups) which, starting with a given group or groups, yield a new group by purely abstract manipulation. Sometimes the constructed group will be isomorphic to a familiar group, but sometimes it will be genuinely new. The point of constructing such groups is that we can immediately apply to them all the theorems about groups that we have proved. Since the new groups are closely related to the ones from which they were built we can then often deduce facts about the original groups which would be more difficult to obtain by other methods. If the constructions seem highly abstract and conceptually difficult at first, the student should persevere until he masters them. He will find that the effort is amply rewarded in later chapters.

The easier of the two constructions is the product. We recall that if A and B are sets then the product set $A \times B$ has as its members all pairs (a, b), where $a \in A$ and $b \in B$. If A and B happen to be groups, then the set $A \times B$ acquires a group structure in a natural way. We write both groups multiplicatively and we define a multiplication on $A \times B$ by the rule:

$$(a, b) \cdot (a', b') = (aa', bb').$$

Since $a, a' \in A$ and $b, b' \in B$ we have $(aa', bb') \in A \times B$, so this multiplication is a binary operation on $A \times B$ and it is clearly associative, because the group multiplications in A and B are both associative. (Write out a complete proof of the associative law in $A \times B$.) If we denote by e_A, e_B the neutral elements of A and B, then $(e_A, e_B) \cdot (a, b) = (e_A a, e_B b) = (a, b)$, and similarly $(a, b) \cdot (e_A, e_B) = (a, b)$, so (e_A, e_B) acts as a neutral element in $A \times B$. Finally, if a^{-1} and b^{-1} denote the inverses of a and b in A and B, respectively, then $(a, b) \cdot (a^{-1}, b^{-1}) = (a^{-1}, b^{-1}) \cdot (a, b) = (e_A, e_B)$, the neutral element of $A \times B$. Thus every element (a, b) has an inverse in $A \times B$, and $A \times B$ is a group, as claimed. There is usually no harm in writing (e, e) for the neutral element. If A and B are written additively it is usual to write $A \times B$ additively too. In this case $(a, b) + (a', b') = (a + a', b + b')$, $-(a, b) = (-a, -b)$, and the neutral element is $(0, 0)$.

This construction can be extended to any number of factors. If $A_1, A_2, \ldots, A_n$ are multiplicative groups then the set $A_1 \times A_2 \times \ldots \times A_n$ has as members all n-tuples $(a_1, a_2, \ldots, a_n)$ with $a_i \in A_i$ and we define multiplication of n-tuples by

$$(a_1, a_2, \ldots, a_n) \cdot (a_1', a_2', \ldots, a_n') = (a_1 a_1', a_2 a_2', \ldots, a_n a_n').$$

The same arguments show that $A_1 \times A_2 \times \ldots \times A_n$ is then a group. Products of groups are associative in the sense that the groups $A \times (B \times C)$, $(A \times B) \times C$ and $A \times B \times C$ are all isomorphic by obvious maps, and we shall not bother to distinguish between them.

Example 6.1. Let $A = B = \mathbf{R}$, the additive group of real numbers. Then $\mathbf{R}^2 = \mathbf{R} \times \mathbf{R}$ is an additive group with $(x, y) + (x', y') = (x + x', y + y')$ for real numbers x, y, x', y'. (This is sometimes known as vector addition.) This group is obviously isomorphic to the additive group $\mathbf{C}$. Indeed, one way of *defining* $\mathbf{C}$ is to say that $\mathbf{C}$ is the set of all pairs of real numbers, with addition defined as above. Similarly $\mathbf{R}^n = \mathbf{R} \times \mathbf{R} \times \ldots \times \mathbf{R}$ (n factors) is an additive group.

Example 6.2. The multiplicative group $\mathbf{R}^*$ has two subgroups $\mathbf{R}^{pos}$ and $\{\pm 1\}$ such that each element $x \in \mathbf{R}^*$ is uniquely expressible as a product of a member of $\mathbf{R}^{pos}$ and a member of $\{\pm 1\}$, namely

$$x = |x| \cdot \text{sign}(x), \text{ where sign}(x) = \frac{x}{|x|}.$$

Thus we have a bijection $\mathbf{R}^* \to \mathbf{R}^{pos} \times \{\pm 1\}$ given by $x \mapsto (|x|, \text{sign}(x))$ and it is easy to see that this is an isomorphism of groups, the point being that

$$xx' \mapsto (|xx'|, \text{sign}(xx')) = (|x||x'|, \text{sign}(x) \text{sign}(x'))$$
$$= (|x|, \text{sign}(x)) \cdot (|x'|, \text{sign}(x'))$$

Example 6.3. A similar argument shows that

$$\mathbf{C}^* \cong \mathbf{R}^{pos} \times T,$$

where T is the circle group. The isomorphism is obtained by taking polar coordinates of non-zero complex numbers.

THEOREM 6A. (i) *If A and B are Abelian groups, so is A × B.* (ii) *If A and B are finite groups of order m, n, respectively then A × B is a finite group of order mn.* (iii) *If A', B' are subgroups of A and B, respectively, then A' × B' is a subgroup of A × B.*

Proof. (i) If A and B are Abelian then

$$(a, b) \cdot (a', b') = (aa', bb') = (a'a, b'b) = (a', b') \cdot (a, b).$$

(ii) The number of distinct choices of an element $a \in A$ and an element $b \in B$ is mn.

(iii) $A' \times B'$ is a subset of $A \times B$ and it contains (e, e). If (a_1, b_1) and $(a_2, b_2) \in A' \times B'$ then $(a_1 a_2, b_1 b_2) \in A' \times B'$ and $(a_1^{-1}, b_1^{-1}) \in A' \times B'$.

Example 6.4. **Z** is an additive subgroup of **R**, so $\mathbf{Z}^2 = \mathbf{Z} \times \mathbf{Z}$ is a subgroup of $\mathbf{R}^2$. It consists of all points in the plane with integer coordinates. If we consider it as a subgroup of $\mathbf{C} \cong \mathbf{R}^2$, it consists of all complex numbers $a + bi$ with $a,b \in \mathbf{Z}$. These complex numbers are known as the *Gaussian integers*.

THEOREM 6B. *Let A,B be finite cyclic groups of order m, n, respectively where m and n are coprime. Then the group C = A × B is cyclic of order mn. The element c = (a, b) is a generator of C if and only if a is a generator of A and b is a generator of B.*

Proof. C certainly has order mn, so in order to show that it is cyclic it is enough to find an element $c = (a, b)$ of order mn. Let a, b be generators of A, B. Then a has order m and b has order n. If $c = (a, b)$ has order r, then $(a, b)^r = (e, e)$, i.e., $a^r = e$ and $b^r = e$, so $m \mid r$ and $n \mid r$ (see Theorem 4D). But m and n are coprime, so $mn \mid r$ by Theorem 5B(v). But r cannot exceed mn, so $r = mn$ and c generates C. Conversely, if $a \in A$, $b \in B$ and if $c = (a, b)$ generates C, then every element of $A \times B$ must be of the form $c^k = (a^k, b^k)$ for some k. It follows that a generates A and b generates B.

Example 6.5. It is not true in general that a product of cyclic groups is cyclic. For example, $\mathbf{Z} \times \mathbf{Z}$ is not cyclic, because the additive powers of a given element (r, s) are the elements (nr, ns) for varying n, and the integers nr, ns are in the same ratio to each other for all these powers. Geometrically, $\mathbf{Z} \times \mathbf{Z}$ is the group of points in the plane with integer coordinates and its cyclic subgroups are the sets of integral points lying on straight lines through the origin. The reader will have seen evidence of these subgroups in a well-planted orchard. Another example is the product $A \times B$ of two cyclic groups of order 2. In this group $(a, b)^2 = (a^2, b^2) = (e, e)$, so every element has order 1 or 2, and there is no element of order 4 to generate the group. This group is known as the *Klein 4-group*.

The construction of quotient groups is more tricky. We start with a group G and a quotient set of G. We then try to use the group structure on G to define a group structure on the quotient set. This cannot always be done, so to analyse the situation we first discuss in general terms the idea of operations on quotient sets.

Let S be a set and $\sim$ an equivalence relation on S. The quotient set $S/\sim$ has as its elements the equivalence classes of $\sim$, and we use the notation $\langle x \rangle$ for the equivalence class containing x. Then $\langle x \rangle = \langle y \rangle \Leftrightarrow x \sim y$. Now suppose that we are given an operation on S; for definiteness we will suppose that it is a binary operation $*$, but the principle is the same for any operation. To define a corresponding operation on $S/\sim$, also to be denoted by $*$, we have to define $\langle x \rangle * \langle y \rangle \in S/\sim$ for every pair of equivalence classes $\langle x \rangle, \langle y \rangle$, and nothing could be more natural than to define it by the equation

$$\langle x \rangle * \langle y \rangle = \langle x * y \rangle. \tag{A}$$

The right-hand side is a member of $S/\sim$ and is uniquely determined if we know x and y. However, *what we are given is not x and y, but $\langle x \rangle$ and $\langle y \rangle$. Since $\langle x \rangle$ and $\langle y \rangle$ do not determine x and y uniquely, equation* (A) *will not in general define an operation on $S/\sim$*. For the right-hand side of (A) to be uniquely determined by $\langle x \rangle$ and $\langle y \rangle$ it is necessary and sufficient that $\langle x * y \rangle$ should be independent of the particular representatives x and y of the two classes $\langle x \rangle$ and $\langle y \rangle$. In other words, if we take new representatives of the same classes, say $\langle x \rangle = \langle x' \rangle$ and $\langle y \rangle = \langle y' \rangle$, the class $\langle x' * y' \rangle$ should be the same as the class $\langle x * y \rangle$.

Similarly, if $\dagger$ is a unary operation on S, the equation

$$\langle x \rangle^\dagger = \langle x^\dagger \rangle \tag{B}$$

will define a unary operation $\dagger$ on $S/\sim$ if and only if $\langle x_1 \rangle = \langle x_2 \rangle \Rightarrow \langle x_1^\dagger \rangle = \langle x_2^\dagger \rangle$ for all $x_1, x_2 \in S$. This condition says that the map $x \mapsto x^\dagger$ from S to S maps each equivalence class into an equivalence class, a condition which obviously is not always satisfied by a given operation and a given equivalence relation.

These important criteria are worth stating as a theorem. We translate them slightly by using the fact that $\langle x_1 \rangle = \langle x_2 \rangle \Leftrightarrow x_1 \sim x_2$.

THEOREM 6C. *Let $\sim$ be an equivalence relation on a set S, with equivalence classes $\langle x \rangle$. Let $\dagger$ and $*$ be, respectively, a unary and a binary operation on S. Then:*

(i) *a necessary and sufficient condition for the equation $\langle x \rangle^\dagger = \langle x^\dagger \rangle$ to define a unary operation on $S/\sim$ is that $x_1 \sim x_2 \Rightarrow x_1^\dagger \sim x_2^\dagger$ for all $x_1, x_2 \in S$.*

(ii) *a necessary and sufficient condition for the equation $\langle x \rangle * \langle y \rangle = \langle x * y \rangle$ to define a binary operation on $S/\sim$ is that $(x_1 \sim x_2$ and $y_1 \sim y_2) \Rightarrow (x_1 * y_1) \sim (x_2 * y_2)$ for all $x_1, x_2, y_1, y_2 \in S$.*

Example 6.6. Let $S = \mathbf{Z}$ and let $\sim$ be congruence modulo 3 (see Theorem 3D). Then $S/\sim$ has three elements $\langle 0 \rangle$, $\langle 1 \rangle$ and $\langle 2 \rangle$. On S we have two binary operations $+$ and $\cdot$ and it is easy to see that if $x_1 \equiv x_2 \pmod 3$ and $y_1 \equiv y_2 \pmod 3$ then $x_1 + y_1 \equiv x_2 + y_2 \pmod 3$ and $x_1 \cdot y_1 \equiv x_2 \cdot y_2 \pmod 3$. Thus $+$ and $\cdot$ do induce operations on the quotient set $S/\sim$. On the other hand, the unary operation $x \mapsto |x|$ on S does not induce a unary operation on $S/\sim$ because, for example, -1 and 2 are in the same class $\langle 2 \rangle$, but their absolute values 1 and 2 are in different classes.

Example 6.7. Again let $S = \mathbf{Z}$ but this time let $x \sim y$ mean that $x = \lambda y$ for some real positive λ. There are three equivalence classes P, N, O, where P and N consist respectively of all positive and all negative integers, and $O = \{0\}$. The unary operation $x \mapsto |x|$ clearly now induces a unary operation on the quotient set, with $|P| = P$, $|N| = P$ and $|O| = O$. Similarly, multiplication on $\mathbf{Z}$ induces a multiplication on $\mathbf{Z}/\sim$ with $P \cdot P = P$, $N \cdot P = N$, $N \cdot N = P$, etc. But addition does not induce an operation on $S/\sim$. (Addition is defined unambiguously for some pairs, e.g. $P + P = P$, $N + N = N$, $O + N = O$, $O + P = P$, but the sum $N + P$ is not defined because the sum of a negative and a positive integer is sometimes positive, sometimes negative, and sometimes 0.)

Suppose now that G is an Abelian group and H is a subgroup of G. We know from Chapter 4 that the cosets of H in G form a partition of G corresponding to the equivalence relation defined by $x \sim y \Leftrightarrow xy^{-1} \in H$. The resulting quotient set, whose members are all the cosets Hx of H, is denoted by G/H instead of $G/\sim$. If we apply the above criteria to this quotient set and the group operations on G we find that G/H does have corresponding operations and is in fact a group with respect to them.

THEOREM 6D. *Let H be a subgroup of an Abelian group G. Then the equation*

$$(Hx)(Hy) = H(xy)$$

defines a binary operation on the set G/H of all cosets of H in G. With respect to this operation G/H is an Abelian group. Its neutral element is $He = H$ and the inverse of Hx is $H(x^{-1})$.

Proof. To see that the equation defines an operation we apply Theorem 6C. We have to verify that if $x_1 \sim x_2$ and $y_1 \sim y_2$ then $x_1 y_1 \sim x_2 y_2$, where $x_1 \sim x_2$ means $x_1 x_2^{-1} \in H$. But if $x_1 x_2^{-1} \in H$ and $y_1 y_2^{-1} \in H$, then their product $x_1 x_2^{-1} y_1 y_2^{-1}$ is in H. Since G is Abelian this implies that $x_1 y_1 y_2^{-1} x_2^{-1} \in H$, i.e., $x_1 y_1 (x_2 y_2)^{-1} \in H$, and this says that $x_1 y_1 \sim x_2 y_2$ as required. Alternatively, we may argue directly

with the cosets and show that the product of an arbitrary element of
Hx and an arbitrary element of Hy always lies in the coset $H(xy)$. To
see this, let $a \in Hx$, $b \in Hy$. Then $a = h_1 x$, $b = h_2 y$ where $h_1, h_2 \in H$.
Hence $ab = h_1 x h_2 y = h_1 h_2 xy$, because G is Abelian. But $h_1 h_2 = h \in H$,
so $ab = hxy \in Hxy$. It now remains to check the group axioms and the
commutative law for G/H. First, it is clear that multiplication is
commutative, because $(Hx)(Hy) = H(xy) = H(yx) = (Hy)(Hx)$. Similarly
the associative law in G implies the associative law in G/H. (Write this
out!) The coset $H = He$ acts as neutral element because
$(He)(Hx) = H(ex) = Hx$ and $(Hx)(He) = H(xe) = Hx$. Finally $H(x^{-1})$ is
inverse to Hx because $(Hx)(H(x^{-1})) = H(xx^{-1}) = He$, and
$(H(x^{-1}))Hx = H(x^{-1}x) = He$. Thus G/H is an Abelian group, and
we call it a *quotient group of G*.

Example 6.8. Let G be the additive group $\mathbf{Z}$ and let $H = n\mathbf{Z}$. Then the
elements of $G/H = \mathbf{Z}/n\mathbf{Z}$ are the residue classes modulo n, so the
quotient set $\mathbf{Z}/n\mathbf{Z}$ is what we have previously called $\mathbf{Z}_n$. It now carries a
group structure which is written additively. If $\langle x \rangle = n\mathbf{Z} + x$ is the residue
class containing x, then addition of residue classes is given by $\langle x \rangle + \langle y \rangle =
\langle x + y \rangle$. For example, the elements of $\mathbf{Z}/6\mathbf{Z}$ are $\langle 0 \rangle, \langle 1 \rangle, \langle 2 \rangle, \langle 3 \rangle, \langle 4 \rangle, \langle 5 \rangle$ and
addition is given by "addition modulo 6"; thus $\langle 2 \rangle + \langle 3 \rangle =
\langle 1 \rangle + \langle 4 \rangle = \langle 5 \rangle$, $\langle 2 \rangle + \langle 4 \rangle = \langle 0 \rangle$, $\langle 3 \rangle + \langle 4 \rangle = \langle 1 \rangle$, etc. Note that
$\langle x \rangle + \langle y \rangle = \langle z \rangle \Leftrightarrow x + y \equiv z \pmod{n}$, and $\langle x \rangle = -\langle y \rangle \Leftrightarrow x \equiv -y \pmod{n}$,
so congruences can always be translated into group equations and vice
versa. One simple consequence of this is that, because in any additive
group the equation $p + x = q$ has a unique solution, the congruence
$p + x \equiv q \pmod{n}$ has as its solutions all members of a single residue
class mod n.

Example 6.9. Let G be the multiplicative group $\mathbf{C}^*$ and let H be the
subgroup T (the circle group). The cosets of T are all the circles centre
0 in the Argand diagram, and there is exactly one corresponding to each
positive real number λ, namely, the circle $T\lambda$ of radius λ. In the
quotient group $\mathbf{C}^*/T$ the product of two circles $T\lambda$, $T\mu$ is given by
$(T\lambda)(T\mu) = T(\lambda\mu)$, that is, the circle whose radius is the product of the
two radii. Thus multiplication in $\mathbf{C}^*/T$ exactly copies multiplication of
positive real numbers and we have an isomorphism of groups
$\mathbf{C}^*/T \cong \mathbf{R}^{pos}$.

Example 6.10. Let G be the multiplicative group $\mathbf{R}^*$ and let H be the
subgroup $\mathbf{R}^{pos}$. Then there are two cosets of H: the set P of positive
reals and the set N of negative reals. The quotient group is a cyclic
group of order 2 with neutral element P and with $N^2 = P$.

Now suppose that G is an arbitrary group and H is a subgroup. We
must distinguish between the left and right cosets of H, so for

definiteness we will define G/H to be the set of *right* cosets Hx of H in G. The quotient set G/H is *not* in general a group, and we have to put conditions on H to make G/H a group. The point is that the equation $(Hx)(Hy) = Hxy$ does not in general define an operation on G/H. If $a \in Hx$ and $b \in Hy$, then $a = h_1 x$, $b = h_2 y$ with $h_1, h_2 \in H$, and $ab = h_1 x h_2 y$, which will not always belong to Hxy. Now we can write $ab = h_1 x h_2 x^{-1} xy$ and this will belong to Hxy if and only if $h_1 x h_2 x^{-1} \in H$. Since $h_1 \in H$, it will be enough to know that $x h_2 x^{-1} \in H$, but we need to know this for every $x \in G$ and every $h_2 \in H$. This motivates the following definition. A subgroup H of G is called a *normal subgroup* of G (written $H \triangleleft G$) if $xhx^{-1} \in H$ for all $x \in G$ and all $h \in H$. We note the following easy consequences of this definition:

(i) In an Abelian group, every subgroup is normal.
(ii) If $H \triangleleft G$ then $x^{-1} hx \in H$ for all $x \in G$ and all $h \in H$.
(iii) If $H \triangleleft G$ then the left and right cosets of H are the same.

(For if $a \in xH$ then $a = xh$ for some $h \in H$, so $ax^{-1} = xhx^{-1} \in H$, which implies $a \in Hx$. Thus $xH \subset Hx$, and similarly $xH \supset Hx$.) We may therefore define G/H to be the set of cosets of H, without further distinction.

THEOREM 6E. *Let G be any group and let H be a normal subgroup of G. Then the equation*

$$(Hx)(Hy) = Hxy$$

defines a binary operation on G/H, and makes G/H a group. Its neutral element is $He = H$ and the inverse of Hx is Hx^{-1}.

Proof. We have just shown that the normality of H is exactly what is needed to prove that the equation defines an operation. The rest of the proof is precisely the same as the proof of Theorem 6D except, of course, that we cannot prove, and do not claim, that G/H is Abelian.

Example 6.11. Let $G = \mathscr{S}_3$ and, in the notation of Examples 4.12 and 4.24, let $H = \{e, a\}$. Then H is not normal because $pap^{-1} = cp^{-1} = cq = b \notin H$. Its left and right cosets are different, as we saw in 4.24. However the subgroup $K = \{e, p, q,\}$ *is* normal (check this!). Its cosets are K and $Ka = \{a, b, c\}$, and the quotient group G/K is cyclic of order two.

Example 6.12. Any group G has two normal subgroups G and $\{e\}$. The corresponding quotients are $G/G \cong \{e\}$ and $G/\{e\} \cong G$.

THEOREM 6F. *Let H be a normal subgroup of a group G. Then*

(i)   *if G is Abelian, so is G/H;*

(ii) *if G is cyclic, so is G/H;*

(iii) *if G is finite, so is G/H and its order divides the order of G;*

(iv) $(Hx)^n$ *is the neutral element in G/H* $\Leftrightarrow x^n \in H;$

(v) *the cosets Hx and Hy commute with each other if and only if* $xyx^{-1}y^{-1} \in H.$

(vi) *G/H is Abelian* $\Leftrightarrow xyx^{-1}y^{-1} \in H$ *for all* $x, y, \in G.$

Proof. (i) follows from Theorem 6D.

(ii) Let G be generated by g. Then any element x of G has the form g^n. Hence any element of G/H has the form $Hx = H(g^n) = (Hg)^n$, so G/H is cyclic, generated by the coset Hg.

(iii) This follows from Lagrange's theorem. If G has order n and H has order m, then $n = rm$, where r is the number of cosets of H, i.e., r is the order of G/H.

(iv) Since $(Hx)^n = Hx^n$, we have

$$(Hx)^n = He \Leftrightarrow Hx^n = He \Leftrightarrow x^n \in He = H.$$

(v) $(Hx)(Hy) = (Hy)(Hx) \Leftrightarrow (Hx)(Hy)(Hx)^{-1}(Hy)^{-1} = He$

$$\Leftrightarrow Hxyx^{-1}y^{-1} = He$$

$$\Leftrightarrow xyx^{-1}y^{-1} \in H.$$

(vi) This follows immediately from (v).

COROLLARY. *Z/nZ is a cyclic group of order n.*

Proof. **Z** is cyclic, generated (additively) by 1, so **Z**/n**Z** is generated by the residue class $\langle 1 \rangle = n\mathbf{Z} + 1$. Of course, this is easy to see directly since the additive powers of $\langle 1 \rangle$ are $\langle 1 \rangle, \langle 2 \rangle, \langle 3 \rangle, \ldots, \langle n \rangle = \langle 0 \rangle$ and these exhaust **Z**/n**Z**.

An important idea which enables us to relate quotient groups with other groups is the idea of homomorphism of groups. It is a generalization of isomorphism and is defined as follows: if A and B are two groups (written multiplicatively), a *homomorphism from A to B* is a function $f : A \to B$ which preserves multiplication, i.e., such that $f(a_1 a_2) = f(a_1) f(a_2)$ for all $a_1, a_2 \in A$. An isomorphism is therefore a homomorphism which is also a bijection. As in the case of an isomorphism, the groups may be additive, or one may be additive and the other multiplicative, etc., in which case the defining equation must be modified accordingly. The essential point is that it should transfer the group operation of A to the group operation of B.

It might be thought that in defining a homomorphism we should insist that the map should preserve inverses and the identity element as well as preserving multiplication. This is unnecessary because the equations $f(e) = e$ and $f(x)^{-1} = f(x^{-1})$ follow from the definition by exactly the same arguments as we gave on p. 50 for isomorphisms.

Example 6.13. We list here some typical homomorphisms which will be referred to later. The reader should in each case make sure he understands why the map is a homomorphism.

(i) Let $f_1 : \mathbf{R}^2 \to \mathbf{R}$ be given by $(x,y) \mapsto x$. Then f_1 is a homomorphism of additive groups since $(x_1, y_1) + (x_2, y_2) = (x_1 + x_2, y_1 + y_2) \mapsto x_1 + x_2$.

(ii) Let $f_2 : \mathbf{C} \to \mathbf{C}$ be given by $z = \bar{z}$ (complex conjugate). Then f_2 is an additive homomorphism since $\overline{z_1 + z_2} = \bar{z}_1 + \bar{z}_2$.

(iii) Let $f_3 : \mathbf{C}^* \to \mathbf{R}^*$ be given by $z \mapsto |z|$. Then f_3 is a multiplicative homomorphism since $|z_1 z_2| = |z_1| |z_2|$.

(iv) Let $f_4 : \mathbf{R} \to \mathbf{R}^*$ be given by $t \mapsto e^t$. Then $s + t \mapsto e^{s+t} = e^s e^t$, so f_4 is a homomorphism from the additive group $\mathbf{R}$ to the multiplicative group $\mathbf{R}^*$.

(v) Let $f_5 : \mathbf{R} \to \mathbf{C}^*$ be given by $t \mapsto e^{2\pi i t}$. Then, as in (iv), f_5 is a homomorphism from additive group to multiplicative group.

(vi) Let $f_6 : GL_n(\mathbf{R}) \to \mathbf{R}^*$ be given by $M \mapsto \det M$, where $\det M$ denotes the determinant of the non-singular matrix M. Since $\det(M_1 M_2) = \det M_1 \cdot \det M_2$, this is a multiplicative homomorphism.

Example 6.14. We can show a connection between homomorphisms and quotient groups immediately. Let $H \triangleleft G$ and consider the quotient map $q : G \to G/H$. For any $x \in G$, $q(x)$ is, by definition, the equivalence class containing x, that is, $q(x) = Hx$. Hence $q(xy) = H(xy) = (Hx) \cdot (Hy) = q(x)q(y)$ and so q is a homomorphism. Indeed one could say that the definition of multiplication in G/H is chosen precisely in order to make the quotient map a homomorphism.

Now consider an arbitrary homomorphism of groups $f : A \to B$. The set of elements $a \in A$ such that $f(a) = e$ is called the *kernel* of f. For an isomorphism this set is just $\{e\}$, but in general it contains other elements. We shall also speak of the *image* of f in the usual set-theoretical sense: $f(A) = \{f(a); a \in A\}$. The next theorem is one of the most basic in abstract algebra. It is called the First Isomorphism Theorem for groups, and it has analogues for other species of algebraic structures. (There are also Second and Third Isomorphism Theorems, but we shall not discuss them in this book.)

THEOREM 6G. *Let A and B be groups and let $f : A \to B$ be a homomorphism. Then*

(i)   *the kernel K of f is a normal subgroup of A;*
(ii)   *the fibres of f are the cosets of K;*

(iii)   the image $C = f(A)$ of f is a subgroup of B;
(iv) $C \cong A/K$.

Proof.   (i) Suppose $k_1, k_2 \in K$ and $a \in A$. Then $f(k_1) = f(k_2) = e$, so $f(k_1 k_2) = f(k_1)f(k_2) = e$ and $f(k_1^{-1}) = f(k_1)^{-1} = e^{-1} = e$. This shows that $k_1 k_2 \in K$ and $k_1^{-1} \in K$. Also $e \in K$ since $f(e) = e$, so K is a subgroup of A. Finally
$$f(ak_1 a^{-1}) = f(a)f(k_1)f(a)^{-1} = f(a)ef(a)^{-1} = f(a)f(a)^{-1} = e,$$
so $ak_1 a^{-1} \in K$ and hence K is normal.

(ii) $f(a_1) = f(a_2) \Leftrightarrow f(a_1)f(a_2)^{-1} = e$
$\Leftrightarrow f(a_1)f(a_2^{-1}) = e$
$\Leftrightarrow f(a_1 a_2^{-1}) = e$
$\Leftrightarrow a_1 a_2^{-1} \in K$
$\Leftrightarrow a_1$ and a_2 are in the same coset of K.

(iii) Since $f(e) = e$, we have $e \in C$. Also, if $c_1, c_2 \in C$ then $c_1 = f(a_1)$, $c_2 = f(a_2)$ for some $a_1, a_2 \in A$; so $c_1 c_2 = f(a_1 a_2)$ and $c_1^{-1} = f(a_1^{-1})$ are both in $f(A) = C$.

(iv) By (ii), f is constant on each coset of K, so we can define a map $f^* : A/K \to C$ by making $f^*(Ka)$ equal to the value of f on each member of Ka. Clearly f^* is surjective, by definition of C, and it is injective by (ii). Finally, *because f is a homomorphism*, we have
$$f^*((Ka_1)(Ka_2)) = f^*(Ka_1 a_2) = f(a_1 a_2) = f(a_1)f(a_2)$$
$$= f^*(Ka_1)f^*(Ka_2).$$

Thus f^* preserves the group operation of A/K and is an isomorphism of groups.

COROLLARY.   *A homomorphism is an injection if and only if its kernel is the trivial subgroup $\{e\}$.*

Proof.   This follows directly from part (ii) of the theorem, because an injection is, by definition, a map whose fibres each consist of a single element.

Example 6.15.   We refer to the six homomorphisms described in Example 6.13 and apply Theorem 6G to each of them.

(i) $f_1 : (x, y) \mapsto x$ from $\mathbf{R}^2$ to $\mathbf{R}$ has image $\mathbf{R}$ and its kernel K is the set of points in $\mathbf{R}^2$ mapping to 0 when projected onto the x-axis; in other words K is the y-axis, and $\mathbf{R}^2/K \cong \mathbf{R}$ by part (iv) of the theorem.

(ii) $f_2 : z \mapsto \bar{z}$ from $\mathbf{C}$ to $\mathbf{C}$ has image $\mathbf{C}$ and kernel $\{0\}$, so the theorem tells us only the trivial fact that $\mathbf{C}/\{0\} \cong \mathbf{C}$.

(iii) $f_3 : z \mapsto |z|$ from $\mathbf{C}^*$ to $\mathbf{R}^*$ has image $\mathbf{R}^{\mathrm{pos}}$ and its kernel is the set $\{z \in \mathbf{C}^*; |z| = 1\} = T$, the circle group. The theorem now tells us that $\mathbf{C}^*/T \cong \mathbf{R}^{\mathrm{pos}}$ (cf. Example 6.9).

(iv) $f_4 : t \mapsto e^t$ from $\mathbf{R}$ to $\mathbf{R}^*$ has image $\mathbf{R}^{\mathrm{pos}}$ and kernel $\{0\}$. It is therefore an injection and induces an isomorphism $\mathbf{R} \to \mathbf{R}^{\mathrm{pos}}$ which we have already met in Example 4.18.

(v) $f_4 : t \mapsto e^{2\pi i t}$ from $\mathbf{R}$ to $\mathbf{C}^*$ is more interesting. Its kernel is the set of real numbers t such that $e^{2\pi i t} = 1$, in other words $\mathbf{Z}$. Its image is the set of all complex numbers of the form $\cos t + i \sin t$ for real t, and this is just the circle group T. The theorem therefore tells us the somewhat surprising fact that $\mathbf{R}/\mathbf{Z} \cong T$. This fact is less surprising if one thinks of taking the real line and wrapping it round a circle of circumference 1 so that all the integers coincide. Each coset of $\mathbf{Z}$ becomes a single point of the circle and addition in $\mathbf{R}$ becomes, after expanding the circle to unit radius, multiplication in T.

(vi) $f_6 : M \mapsto \det M$ from $\mathrm{GL}_n(\mathbf{R})$ to $\mathbf{R}^*$ is a surjection since, given $\alpha \in \mathbf{R}^*$, the diagonal matrix with diagonal entries $\alpha, 1, 1, \ldots, 1$ is non-singular and has determinant α. The kernel of f_6 is, by definition, the special linear group $\mathrm{SL}_n(\mathbf{R})$, so we have $\mathrm{SL}_n(\mathbf{R}) \triangleleft \mathrm{GL}_n(\mathbf{R})$ and $\mathrm{GL}_n(\mathbf{R})/\mathrm{SL}_n(\mathbf{R}) \cong \mathbf{R}^*$.

Example 6.16. Let G be any cyclic group, generated by an element g. Let $f : \mathbf{Z} \to G$ be the map defined by $r \mapsto g^r$. Since $g^{r+s} = g^r g^s$, f is a homomorphism from the additive group $\mathbf{Z}$ to G. Its image is G since G is generated by g. Hence $G \cong \mathbf{Z}/K$ where K is the kernel of f. The possible kernels are $\{0\}$, if g has infinite order, or $n\mathbf{Z}$, if g has order n. Hence any cyclic group is isomorphic to $\mathbf{Z}$ or to $\mathbf{Z}/n\mathbf{Z}$ for some $n \geq 1$. This gives an alternative proof of Theorem 4E and its Corollary.

The universal property of quotient sets described in Theorem 2E has an analogue for quotient groups which we now prove. It is essentially another way of stating Theorem 6G in a slightly stronger form.

THEOREM 6H. *Let A be a group, let $N \triangleleft A$, and let $q : A \to A/N$ be the quotient homomorphism. If $f : A \to B$ is any homomorphism of groups such that $f(N) = \{e\}$, then there is a unique homomorphism $f^* : A/N \to B$ such that $f = f^* \circ q$.*

Proof. The arguments are very similar to those of Theorem 6G. If a_1, a_2 are in the same coset of N in A then $a_1 a_2^{-1} \in N$, so $f(a_1 a_2^{-1}) = e$, which implies $f(a_1) = f(a_2)$. Thus f is constant on the cosets of N and we may define a map $f^* : A/N \to B$ by the rule $f^*(Na) = f(a)$. (Compare this with Theorem 2E.) Clearly $f = f^* \circ q$ since $f^*(q(a)) = f^*(Na) = f(a)$. That f^* is a homomorphism follows from the definition of multiplication in A/N; for

$$f^*((Na_1)(Na_2)) = f^*(Na_1a_2) = f(a_1a_2) = f(a_1)f(a_2)$$
$$= f^*(Na_1)f^*(Na_2).$$

That f^* is unique follows from the equation $f = f^* \circ q$; for if $f = g \circ q$ then $g(q(a)) = f(a)$ for every $a \in A$, so $g(Na) = f(a)$ and $g = f^*$. The converse of this theorem is of course trivial, that is, if f^* is a homomorphism from A/N to B and if $f = f^* \circ q$, then $f(N) = \{e\}$.

Example 6.17. Let $q : \mathbf{Z} \to \mathbf{Z}/n\mathbf{Z}$ be the quotient homomorphism which sends each integer to its residue class mod n. We denote the class of x by $\langle x \rangle_n$. Now let $f : \mathbf{Z} \to B$ be any group homomorphism such that $f(n) = e$. Then the kernel of f must contain the subgroup $n\mathbf{Z}$ generated by n, so we may apply Theorem 6H to obtain a homomorphism $f^* : \mathbf{Z}/n\mathbf{Z} \to B$ such that $f^*(\langle x \rangle_n) = f(x)$ for all $x \in \mathbf{Z}$. In particular, if f is also a quotient homomorphism from $\mathbf{Z}$ to $\mathbf{Z}/m\mathbf{Z}$, say, then $f(n) = \langle n \rangle_m = 0$ if and only if $m \mid n$. Thus, when $m \mid n$, we have a group homomorphism $\mathbf{Z}/n\mathbf{Z} \to \mathbf{Z}/m\mathbf{Z}$ which sends $\langle x \rangle_n$ to $\langle x \rangle_m$ for every integer x. If $m \nmid n$, no such map exists, as the reader should convince himself.

Exercises

1. Let A and B be groups. Let $a \in A$, $b \in B$ be elements of orders m, n, respectively. Prove that the element (a, b) of $A \times B$ has order N, where N is the least common multiple of m and n.
2. Prove the converse of Theorem 6B, namely, if A and B are cyclic groups and if $A \times B$ is also cyclic, then A and B are finite and their orders are coprime.
3. Let $f : A \to B$ be a group homomorphism and suppose that f is a surjection. Prove:

    (i) if A is Abelian, so is B;

    (ii) if A is cyclic, so is B;

    (iii) if B is *torsion-free*, so is A. (A group is torsion-free if the only element of finite order is e.)

4. Prove that if G is a group of order $2n$ and H is a subgroup of order n then H is normal.
5. Prove that $(\mathbf{R} \times \mathbf{R})/(\mathbf{Z} \times \mathbf{Z}) \cong T \times T$.
6. Prove that $T/\{\pm 1\} \cong T$. What about T/P_n, where P_n is the group of nth roots of 1?
7. Prove that $\mathbf{C}^*/\mathbf{R}^{pos} \cong T$. (Hint: find a homomorphism $\mathbf{C}^* \to T$ which sends $\mathbf{R}^{pos}$ to 1.)

8. Let G be any group and let H be the set of all elements of G which can be expressed as a product of squares of elements of G. Prove that H is a normal subgroup of G and that G/H is Abelian.
9. Let A and B be groups and let $G = A \times B$. Show that the set $H = \{(a, e); a \in A\}$ is a normal subgroup of G and establish the isomorphisms: $H \cong A$; $G/H \cong B$.
10. Prove that the group $\mathbf{Q}/\mathbf{Z}$ is torsion-free. (See Exercise 3 above for the definition.)
11. Prove that the set

$$P = \bigcup_{n=1}^{\infty} P_n$$

of *all* roots of 1 in $\mathbf{C}$ is a subgroup of the circle group. Prove that $\mathbf{Q}/\mathbf{Z} \cong P$.
12. Prove that any group of order 4 is either cyclic or is isomorphic to the product of two cyclic groups of order 2.

CHAPTER 7

Linear Congruences in Z

For any positive integer n, congruence modulo n is an equivalence relation on $\mathbf{Z}$ whose classes are the residue classes modulo n. These classes are the elements of the quotient group $\mathbf{Z}/n\mathbf{Z}$ which is cyclic of order n, the group operation being "addition modulo n". We have already pointed out that statements of congruence between integers can always be translated into statements of equality between residue classes in the group $\mathbf{Z}/n\mathbf{Z}$, so it is not surprising that many of the rules for manipulating equations are also valid for congruences. We now make this more precise and point out the pitfalls — some standard procedures for solving equations are not applicable to congruences.

We first take a fixed modulus n and consider arbitrary congruences of the form $a \equiv b \pmod{n}$, where a and b may stand for complicated expressions involving integer variables, the values of a and b being always integers. We have the following rules, as for equations:

(i) $a \equiv b \pmod{n} \Leftrightarrow a - b \equiv 0 \pmod{n}$;

(ii) $a \equiv b + c \pmod{n} \Leftrightarrow a - c \equiv b \pmod{n}$;

(iii) $a \equiv b \pmod{n} \Leftrightarrow -a \equiv -b \pmod{n}$.

All three follow from the corresponding rules for equations in $\mathbf{Z}/n\mathbf{Z}$, or can be proved directly from the definition.

(iv) $a \equiv b \pmod{n} \Rightarrow ra \equiv rb \pmod{n}$ for all integers r.

Proof. If $n \mid (a - b)$ then $n \mid ra - rb = r(a - b)$. Alternatively, in $\mathbf{Z}/n\mathbf{Z}$, if $\langle a \rangle = \langle b \rangle$ then the rth additive powers of $\langle a \rangle$ and $\langle b \rangle$ are equal, i.e., $r\langle a \rangle = r\langle b \rangle$. Since $r\langle a \rangle = \langle ra \rangle$ this implies $\langle ra \rangle = \langle rb \rangle$.
Warning: This implication only goes one way. If we were dealing with *equations* in $\mathbf{Z}$ then $ra = rb$ would imply $a = b$ (cancellation law) provided that $r \neq 0$. However, if $ra \equiv rb \pmod{n}$ we cannot conclude that $a \equiv b \pmod{n}$ even if $r \not\equiv 0 \pmod{n}$. For example $2.4 \equiv 2.1 \pmod 6$ and $2 \not\equiv 0 \pmod 6$, but we cannot cancel the 2 and conclude that $4 \equiv 1 \pmod 6$. In group-theoretical terms, what we are saying here is that in $\mathbf{Z}/n\mathbf{Z}$, $r\langle a \rangle = r\langle b \rangle \not\Rightarrow \langle a \rangle = \langle b \rangle$. If we translate this into multiplicative notation for a group G it says that if $x, y \in G$ and $x^r = y^r$ in G, then x and y need not be equal, even in a cyclic group.

Of course, sometimes r can be cancelled from a congruence $ra \equiv rb \pmod{n}$ and it is important to have a usable criterion. Also, a new phenomenon appears here: sometimes one can cancel r *at the expense of changing the modulus*. The main questions in this direction are settled by the following theorem.

THEOREM 7A. *Let n be a positive integer and let a,b,r be integers.*
(i) *If $(r, n) = 1$, then $ra \equiv rb \pmod{n} \Leftrightarrow a \equiv b \pmod{n}$.*
(ii) *If $r \mid n$, say $n = rn'$, then $ra \equiv rb \pmod{n} \Leftrightarrow a \equiv b \pmod{n'}$.*
(iii) *In general, if $r \neq 0$, let $(r, n) = d$. Then $ra \equiv rb \pmod{n} \Leftrightarrow a \equiv b \pmod{n_1}$, where $n = n_1 d$.*

Proof. (i) If $ra \equiv rb \pmod{n}$ then $n \mid r(a - b)$. If $(n, r) = 1$ we may conclude that $n \mid (a - b)$, by Theorem 5B, i.e., $a \equiv b \pmod{n}$. The converse implication has already been proved and is true for any r.

(ii) Let $n = rn'$. If $a \equiv b \pmod{n'}$, then $n' \mid a - b$, so clearly $rn' \mid ra - rb$, i.e., $ra \equiv rb \pmod{n}$. Conversely, if $ra \equiv rb \pmod{n}$, then $rn' \mid ra - rb$, so $r(a - b) = rn'k$ for some integer k. We may now apply the cancellation law in $\mathbf{Z}$ to deduce that $a - b = n'k$. (Note that $r \neq 0$ since $n = rn'$ is positive.) Hence $a \equiv b \pmod{n'}$.

(iii) Let $r \neq 0$ and let $(r, n) = d$. Then $d \neq 0$ and $r = r_1 d$, $n = n_1 d$, where $(r_1, n_1) = 1$ (Theorem 5B). If $ra \equiv rb \pmod{n}$ then $dr_1 a \equiv dr_1 b \pmod{dn_1}$ and, applying part (ii) with r replaced by d, we deduce that $r_1 a \equiv r_1 b \pmod{n_1}$. Since $(r_1, n_1) = 1$, this implies $a \equiv b \pmod{n_1}$, by (i).

Example 7.1. Part (iii) of this theorem is important theoretically and it includes the other two parts as special cases. In practice, part (ii) is easier to use if the numbers involved are small, and is usually all one needs. For example, to solve the congruence $198x \equiv 132 \pmod{110}$, (that is, to find all integers x for which it is true) we look first for common divisors of 198, 132 and 110, and apply part (ii). We obtain
$$198x \equiv 132 \pmod{110} \Leftrightarrow 99x \equiv 66 \pmod{55}$$
$$\Leftrightarrow 9x \equiv 6 \pmod{5}.$$

We could now simplify further by using part (i) with $r = 3$. Since 3 is coprime to 5 we have $9x \equiv 6 \pmod 5 \Leftrightarrow 3x \equiv 2 \pmod 5$, which looks better but is no easier to solve. Various other methods of reduction are available and may be more practical than this last one. For instance, since $9 \equiv -1$ and $6 \equiv 1 \pmod 5$ we see that
$$9x \equiv 6 \pmod 5 \Leftrightarrow -x \equiv 1 \pmod 5 \Leftrightarrow x \equiv -1 \pmod 5,$$
and this solves the congruence completely: the solutions are all integers of the form $x = 5m - 1$. Of course, if we happen to observe that $6 \equiv -9 \pmod 5$ we can argue instead that
$9x \equiv 6 \pmod 5 \Leftrightarrow 9x \equiv -9 \pmod 5$ and then use part (i) of the theorem

to deduce that $9x \equiv -9 \pmod 5 \Leftrightarrow x \equiv -1 \pmod 5$, since $(9, 5) = 1$. What is needed is a systematic procedure to replace these *ad hoc* methods when the numbers are less amenable. This is our next aim.

A linear congruence $px \equiv q \pmod n$ may not have any solutions; for example $2x \equiv 3 \pmod 4$ cannot be true for any integer x since $2x - 3$ is always odd. On the other hand, if $px \equiv q \pmod n$ has one solution it has infinitely many since if $px \equiv q \pmod n$ is true for a particular x it is obviously also true for any integer congruent to $x \pmod n$. In fact this shows that the solutions of $px \equiv q \pmod n$, if any, form a union of residue classes modulo n. Taking this fact into account we shall say that the linear congruence $px \equiv q \pmod n$ *has k solutions modulo n* if its solutions comprise exactly k complete residue classes modulo n. That k may be more than 1 can be seen from the congruence $2x \equiv 2 \pmod 4$ which is satisfied not only by all $x \equiv 1 \pmod 4$ but by all $x \equiv 3 \pmod 4$. It has two solutions modulo 4.

THEOREM 7B. (i) *For given p, n the congruence $px \equiv 1 \pmod n$ has a solution if and only if $(p, n) = 1$. If $(p, n) = 1$, there is a unique solution modulo n.* (ii) *For given p, q, n the congruence $px \equiv q \pmod n$ has a solution if and only if $d \mid q$, where $d = (p, n)$. If $d \mid q$ then there are exactly d solutions modulo n.*

Proof. (i) is a special case of (ii), but it is worth giving the simple argument in this case. The congruence $px \equiv 1 \pmod n$ has a solution if and only if there exist integers x and y such that $px + ny = 1$. But such integers exist if and only if $(p, n) = 1$, by Theorem 5B (iii). Assuming that $(p, n) = 1$ it is easy to see that the solution is unique modulo n, for if $px \equiv 1 \pmod n$ and $px' \equiv 1 \pmod n$ then $px \equiv px' \pmod n$ and therefore $x \equiv x' \pmod n$ by Theorem 7A (i).

(ii) The argument is similar but a bit harder. The congruence $px \equiv q \pmod n$ has a solution $\Leftrightarrow (\exists x, y \in \mathbf{Z})(px + ny = q) \Leftrightarrow (p, n) \mid q$, by the Corollary to Theorem 5A. Assuming that $d = (p, n) \mid q$, we may write $p = dp_1$, $n = dn_1$, $q = dq_1$ and then by Theorem 7A (ii), $px \equiv q \pmod n \Leftrightarrow p_1 x \equiv q_1 \pmod{n_1}$. But $(p_1, n_1) = 1$ (by Theorem 5B (ii)) so if x and x' are any two solutions then $p_1 x \equiv p_1 x' \pmod{n_1}$ and hence $x \equiv x' \pmod{n_1}$ by Theorem 7A (i). Thus, in this case the congruence has a unique solution modulo n_1. It remains to show that every residue class modulo n_1 comprises d residue classes modulo $n = dn_1$. So let R be a residue class modulo n_1. Clearly, R is a union of residue classes modulo n, because $x \equiv y \pmod n \Rightarrow x \equiv y \pmod{n_1}$. So we simply have to count the number of integers in the range 0, 1, 2, ..., $n - 1$ which lie in R. Let r be the unique member of R in the range $0 \leqslant r < n_1$. Then the required integers are r, $r + n_1$, $r + 2n_1, \ldots, r + (d - 1)n_1$, and there are exactly d of them. (The next candidate $r + dn_1 = r + n$ is too big.)

We are now in a position to solve any linear congruence $px \equiv q \pmod{n}$. First compute $(p, n) = d$ by Euclid's algorithm. If $d \nmid q$ there are no solutions. If $d \mid q$, use the algorithm to express q in the form $q = px + ny$. This gives one solution x, and the complete set of solutions consists of all integers congruent to this x modulo n_1, where $n = dn_1$. For small values of p, q, n short-cuts may be possible as in Example 7.1.

Example 7.2. To solve $27x \equiv 13 \pmod{60}$, observe that $(27,60) = 3$ and $3 \nmid 13$, so there are no solutions. To solve $27x \equiv 15 \pmod{60}$, since now $3 \mid 15$, we argue:

$$27x \equiv 15 \pmod{60} \Leftrightarrow 9x \equiv 5 \pmod{20}$$
$$\Leftrightarrow 9x \equiv 45 \pmod{20}$$
$$\Leftrightarrow x \equiv 5 \pmod{20}.$$

So the solutions modulo 60 are $x \equiv 5, 25, 45 \pmod{60}$.

Example 7.3. To solve $224x \equiv 154 \pmod{385}$, we use Euclid's algorithm:

$$385 = 224.2 - 63$$
$$224 = 63.4 - 28$$
$$63 = 28.2 + 7.$$

Thus $(385,224) = 7$ and $154 = 7.22$, so solutions exist. Also, from the algorithm, $7 = 63 - 28.2 = 224.2 - 63.7 = 385.7 - 224.12$, so $154 = 385.7.22 - 224.12.22$. This gives us one solution $x = -12.22 = -264 \equiv 121 \pmod{385}$. According to Theorem 7B there are 7 solutions modulo 385 and they are equally spaced at intervals of $385/7 = 55$. Hence the solutions are $x \equiv 11, 66, 121, 176, 231, 286, 341 \pmod{385}$.

The solution of simultaneous linear congruences is more difficult. Consider, for example, a pair of congruences in two unknowns

$$\begin{cases} ax + by \equiv c \pmod{n} & (1) \\ a'x + b'y \equiv c' \pmod{n}. & (2) \end{cases}$$

For equations the procedure would be to eliminate x from (2) using (1), solve for y and then use (1) to find x. This procedure can be justified if $a \neq 0$, but for congruences there are snags. We cannot divide by a since we must only work with integers. So the best we could do would be to multiply (1) and (2) by suitable integers until the coefficients of x were equal. However, if we multiply by integers which are not coprime to n we may alter the solutions of the individual congruences and perhaps also their simultaneous solutions. So elimination may not be possible. If n happens to be prime this problem disappears (as we shall see later) and the normal method works. Otherwise one must proceed with extreme caution. It is not profitable to develop a formal theory of such congruences. We confine ourselves to two examples.

Example 7.4. To solve

$$\begin{cases} 3x - 4y \equiv 2 \pmod{7} \\ 5x + 6y \equiv 3 \pmod{7} \end{cases}, \qquad (3)$$

we may multiply the first congruence by 5 and the second by 3 without altering their solutions.

So $(3) \Leftrightarrow \begin{cases} 15x - 20y \equiv 10 \pmod{7} \\ 15x + 18y \equiv 9 \pmod{7} \end{cases}$

$\Leftrightarrow \begin{cases} 15x - 20y \equiv 10 \pmod{7} \\ 38y \equiv -1 \pmod{7} \end{cases}.$

Now $38y \equiv -1 \pmod{7} \Leftrightarrow 3y \equiv -1 \equiv 6 \pmod{7} \Leftrightarrow y \equiv 2 \pmod{7}$. So the solutions of (3) are given by $y \equiv 2$ and $15x - 40 \equiv 10 \pmod{7}$. Hence there is a unique solution $x \equiv 1, y \equiv 2 \pmod{7}$.

Example 7.5. To solve

$$\begin{cases} 4x - 6y \equiv 2 \pmod{30} \\ 5x + 22y \equiv 7 \pmod{30} \end{cases}.$$

We cannot eliminate x without changing the solutions. However, since $(22, 30) = 2$ we can achieve a coefficient 2 for y in the second equation and then eliminate y. To do this we need to solve $22k \equiv 2 \pmod{30}$ and, finding that $k = 11$ is a solution, we multiply by 11, noting that $(11, 30) = 1$.

Thus $\begin{cases} 4x - 6y \equiv 2 \pmod{30} \\ 5x + 22y \equiv 7 \pmod{30} \end{cases} \Leftrightarrow \begin{cases} 4x - 6y \equiv 2 \pmod{30} \\ 25x + 2y \equiv 17 \pmod{30} \end{cases}$

$\Leftrightarrow \begin{cases} 25x + 2y \equiv 17 \pmod{30} \\ 19x \equiv 23 \pmod{30} \end{cases}.$

Note that at the last stage we must not keep the first congruence $4x - 6y \equiv 2$ instead of $25x + 2y \equiv 17$, for then we would not be able to recover $25x + 2y \equiv 17$ after multiplying it by 3. From this point on it is easy to obtain the unique solution $x \equiv -13 \pmod{30}$ for the congruence $19x \equiv 23 \pmod{30}$ and hence the two simultaneous solutions modulo 30 of the original pair of congruences: either $x \equiv -13$ and $y \equiv 6$, or $x \equiv -13$ and $y \equiv -9$.

An alternative method would be to eliminate by the standard procedures for linear equations making sure that at each stage at least the new congruences are implied by the old, if not conversely. At the end of the argument one will have a set of congruences whose solutions contain all solutions of the originals, but possibly some extraneous ones. One must then substitute back to find which ones are genuine. Thus for the same pair of congruences one might argue

$$\begin{cases} 4x - 6y \equiv 2 \pmod{30} \\ 5x + 22y \equiv 7 \pmod{30} \end{cases} \Rightarrow \begin{cases} 20x - 30y \equiv 10 \pmod{30} \\ 20x + 88y \equiv 28 \pmod{30} \end{cases}$$

$$\Rightarrow \begin{cases} 20x \equiv 10 \pmod{30} \\ 28y \equiv 18 \pmod{30} \end{cases}.$$

This implies $y \equiv 6$ or $y \equiv -9 \pmod{30}$ but, unfortunately, gives ten possible values for $x \pmod{30}$, nine of which do not in fact give solutions of the original congruences. Of course, one can eliminate all but two of them immediately by observing that the original congruences also imply (by subtraction) $x + 28y \equiv 5$, and therefore $x \equiv 5 - 28.6$ or $x \equiv 5 + 28.9$. Both these give $x \equiv -13 \pmod{30}$ and we obtain the same solutions as above.

We now turn to simultaneous linear congruences of a different type — several congruences in one unknown but with different moduli. Here we have no analogy with simultaneous equations to guide us, but there is one important special case in which we can solve the problem completely without much difficulty. This is the case in which the different moduli are pairwise coprime. The so-called Chinese Remainder Theorem gives us the necessary information.

THEOREM 7C. *Let m, n be positive integers with $(m, n) = 1$. Let a and b be arbitrary integers. Then the congruences*

$$\begin{cases} x \equiv a \pmod{m} \\ x \equiv b \pmod{n} \end{cases}$$

have a common solution, and the set of common solutions is a single residue class modulo mn.

Proof. Clearly, if x is a common solution and if $x' \equiv x \pmod{mn}$, then x' is also a common solution. Conversely, if x and x' are two common solutions then $m \mid x - x'$ and $n \mid x - x'$, so $mn \mid x - x'$ by Theorem 5B(v), since $(m, n) = 1$. Thus, if there are any common solutions, they form a single residue class modulo mn, and the only problem is the existence of solutions. We give two proofs. The first proof uses the groups $A = \mathbf{Z}/m\mathbf{Z}$ and $B = \mathbf{Z}/n\mathbf{Z}$ which are cyclic of orders m and n respectively. Since $(m, n) = 1$, Theorem 6B tells us that the product group $A \times B$ is cyclic of order mn and is generated (additively) by the element $g = (\langle 1 \rangle_m, \langle 1 \rangle_n)$. Here $\langle 1 \rangle_m$ is the residue class (mod m) containing 1 and is a generator of A; similarly $\langle 1 \rangle_n$ generates B. The (additive) powers of g are

$$rg = (r\langle 1 \rangle_m, r\langle 1 \rangle_n) = (\langle r \rangle_m, \langle r \rangle_n)$$

and these exhaust $A \times B$. Hence, for arbitrary integers a, b, there is an integer r such that

$$(\langle r \rangle_m, \langle r \rangle_n) = (\langle a \rangle_m, \langle b \rangle_n),$$

and this r is the required solution.

This nice argument does not, however, help one to find a solution, so we give a more constructive proof as well. Since $(m, n) = 1$, we can find (by Euclid's algorithm) integers p and q such that $mp + nq = 1$. We look for a solution of the congruences in the special case $a = 1, b = 0$, i.e., we look for x_1 such that $x_1 \equiv 1 \pmod{m}$ and $x_1 \equiv 0 \pmod{n}$. Clearly $x_1 = nq$ is such a solution. Similarly $x_2 = mp$ is a solution of $x_2 \equiv 0 \pmod{m}$ and $x_2 \equiv 1 \pmod{n}$. Having solved these special cases we can combine them to get a solution of the general case, for clearly, if we put $x = ax_1 + bx_2$, we have $x \equiv a \cdot 1 + b \cdot 0 \pmod{m}$ and $x \equiv a \cdot 0 + b \cdot 1 \pmod{n}$, as required.

COROLLARY 1. *Let $n_1, n_2, \ldots, n_r$ be pairwise coprime positive integers. Let $a_1, a_2, \ldots, a_r$ be arbitrary integers. Then the congruences $x \equiv a_i \pmod{n_i}, i = 1, 2, \ldots, r$ have a common solution and the common solutions form a single residue class modulo n, where $n = n_1 n_2 \ldots n_r$.*

Proof. We use induction on r. If $r = 1$ there is nothing to prove. If $r = 2$, the assertion is the same as the theorem. For $r > 2$, we may apply the induction hypothesis to obtain a solution $x = a_0$ of the first $r - 1$ congruences. The set of all such solutions is a residue class modulo $n_0 = n_1 n_2 \ldots n_{r-1}$. Thus the common solutions of all r congruences are precisely the common solutions of $x \equiv a_0 \pmod{n_0}$ and $x \equiv a_r \pmod{n_r}$. The corollary follows by another application of the theorem, provided that we know that $n_0 = n_1 n_2 \ldots n_{r-1}$ is coprime to n_r. This has already been set as an exercise (Chapter 5, Exercise 8) but we give a proof here for completeness. If $(n_0, n_r) \neq 1$ then there is a common factor d of n_0 and n_r which is not a unit. This d is divisible by a prime p, and we have $p \mid n_r$ and $p \mid n_0 = n_1 n_2 \ldots n_{r-1}$. By Theorem 5C(iv), $p \mid n_i$ for some i in the range $1 \leq i \leq r-1$ and hence n_i is not coprime to n_r, contrary to hypothesis. This contradiction proves that $(n_0, n_r) = 1$.

COROLLARY 2. (*The General Chinese Remainder Theorem*). Let $n_1, \ldots, n_r$ be pairwise coprime positive integers. Let $b_i, c_i (i = 1, 2, \ldots, r)$ be arbitrary integers. Then the congruences $b_i x \equiv c_i \pmod{n_i}$ $(i = 1, 2, \ldots, r)$ have a common solution if and only if each of them has a solution.

Proof. If the ith congruence has a solution a_i then any $x \equiv a_i \pmod{n_i}$ is also a solution. If a solution a_i exists for each i, then Corollary 1 guarantees the existence of an x satisfying all the congruences.

Example 7.6. To solve the simultaneous congruences $x \equiv 3 \pmod{10}$, $2x \equiv 5 \pmod 7$ and $6x \equiv 3 \pmod 9$ we first check that they each have solutions and reduce them to the equivalent congruences:

$$x \equiv 3 \pmod{10}, \quad x \equiv -1 \pmod 7, \quad \text{and} \quad x \equiv 2 \pmod 3.$$

The moduli are pairwise coprime, so the common solutions form a single residue class modulo 210, and it is only necessary to discover one solution. The last two congruences obviously have the common solutions $x \equiv -1 \pmod{21}$, so we want a common solution of this and $x \equiv 3 \pmod{10}$. Now $21.1 - 10.2 = 1$ so $x_1 = -20$ and $x_2 = 21$ are solutions of $x_1 \equiv 1 \pmod{21}, x_1 \equiv 0 \pmod{10}, x_2 \equiv 0 \pmod{21}$, $x_2 \equiv 1 \pmod{10}$. Hence $x = (-20).(-1) + 21.3 = 83$ is the solution we seek. The solutions of the original congruences are $x \equiv 83 \pmod{210}$.

The connections between congruences and cyclic groups have already been made plain. The relationship can be exploited in both directions and we shall now undertake a more detailed study of the structure of cyclic groups — their generators, their subgroups and the orders of their elements — in which the theory of congruences will play a major part. The resulting abstract theorems can be used in turn to give information of a number-theoretical nature about integers.

THEOREM 7D. *Let G be a cyclic group of order n, generated by an element g. Then g^s generates G if and only if $(s, n) = 1$.*

Proof. Let $h = g^s$. If h generates G, then g must be a power of h. Conversely, if g is a power of h, then every element of G, being a power of g is also a power of h. Thus h generates G if and only if there is an integer x such that $g = h^x = g^{sx}$. But $g^{sx} = g \Leftrightarrow sx \equiv 1 \pmod n$, since g has order n (see Theorem 4D). This congruence has a solution x if and only if $(s, n) = 1$ (Theorem 7B).

Definition. *Euler's function* ϕ is defined on the set of positive integers by the rule: $\phi(n)$ is the number of integers r in the range $1 \leqslant r \leqslant n$ which are coprime to n. Thus, for example, $\phi(1) = 1$,

$\phi(2) = 1, \phi(3) = 2, \phi(4) = 2, \phi(5) = 4, \phi(6) = 2$. We shall reserve the symbol ϕ for this function from now on.

COROLLARY. *In a cyclic group G of order n, the number of elements which (individually) generate G is $\phi(n)$.*

Proof. The elements of G can be written as g^s for $s = 1, 2, \ldots, n$ and, by the theorem, those which generate G are the ones with $(s, n) = 1$.

Example 7.7. An nth root of 1 in $\mathbb{C}$ is called a *primitive nth root of 1* if it is not a kth root of 1 for any $k < n$, in other words, if its order is exactly n. But this is the same as saying that it generates the cyclic group $\mathcal{P}_n$ of all nth roots of 1. The group $\mathcal{P}_n$ has order n, so the number of primitive nth roots of 1 is $\phi(n)$. The primitive roots are $e^{2\pi i s/n}$, where $1 \leqslant s \leqslant n$ and $(s, n) = 1$.

THEOREM 7E. *If $(m, n) = 1$, then $\phi(mn) = \phi(m)\phi(n)$.*

Proof. There are many proofs of this result, but perhaps the neatest uses the characterisation of $\phi(n)$ as the number of generators in a cyclic group. We let A and B be cyclic groups of orders m and n, respectively. By Theorem 6B, the group $A \times B$ is cyclic of order mn, since $(m, n) = 1$. The generators of $A \times B$ are the elements (a, b), where a generates A and b generates B. Since, by Theorem 7D, there are $\phi(m)$ possible choices of a and $\phi(n)$ choices of b, the number of different generators of $A \times B$ is $\phi(m)\phi(n)$. But this number is $\phi(mn)$, by Theorem 7D again.

COROLLARY 1. *If $n_1, n_2, \ldots, n_r$ are pairwise coprime positive integers, then $\phi(n_1 n_2 \ldots n_r) = \phi(n_1)\phi(n_2) \ldots \phi(n_r)$.*

Proof. This is an easy induction on r, using the theorem and the fact that $n_1 n_2 \ldots n_{r-1}$ is coprime to n_r (cf. the proof of Theorem 7C, Corollary 1.)

COROLLARY 2. *For any $n > 0$,*

$$\phi(n) = n \prod_{p \mid n} \left(1 - \frac{1}{p}\right),$$

where the product is taken over all distinct prime divisors p of n.

Proof. Let $p_1, p_2, \ldots, p_r$ be the distinct prime divisors of n and let $n = p_1^{\alpha_1} p_2^{\alpha_2} \ldots p_r^{\alpha_r}$ ($\alpha_i \geqslant 1$). Then $p_i^{\alpha_i}$ and $p_j^{\alpha_j}$ are coprime if $i \neq j$

(exercise) so

$$\phi(n) = \sum_{i=1}^{r} \phi(p_i^{\alpha_i})$$

by Corollary 1. Now if p is prime then $\phi(p^\alpha)$ ($\alpha \geq 1$) is the number of integers s in the range $1 \leq s \leq p^\alpha$ which are not divisible by p. The number which *are* divisible by p is clearly

$$\frac{1}{p} \cdot p^\alpha, \quad \text{so} \quad \phi(p^\alpha) = p^\alpha \left(1 - \frac{1}{p}\right).$$

Hence

$$\phi(n) = \prod_{i=1}^{r} \phi(p_i^{\alpha_i}) = \prod_{i=1}^{r} p_i^{\alpha_i}\left(1 - \frac{1}{p_i}\right) = n \prod_{i=1}^{r}\left(1 - \frac{1}{p_i}\right).$$

THEOREM 7F. *Let G be a cyclic group of order n, generated by an element g. Then*

(i) *Every subgroup of G is cyclic and has order a divisor of n.*

(ii) *The element g^r has order $n/(n, r)$.*

(iii) *The number of solutions in G of the equation $x^m = e$ is (m, n).*

(iv) *There is exactly one subgroup of order d for each divisor d of n.*

(v) *The number of elements in G of order exactly d is*

$$\begin{cases} \phi(d) & \text{if } d \mid n, \\ 0 & \text{if } d \nmid n. \end{cases}$$

Proof. (i) This has already been proved. See Theorem 4C, Corollary, and Theorem 4F, Corollary 1.

(ii) Let $h = g^r$. Then $h^s = g^{rs} = e \Leftrightarrow n \mid rs$. If $(n, r) = d$ then $n = dn_1$, $r = dr_1$, with $(n_1, r_1) = 1$. Hence $n \mid rs \Leftrightarrow dn_1 \mid dr_1 s \Leftrightarrow n_1 \mid r_1 s \Leftrightarrow n_1 \mid s$, and it follows that the order of h is $n_1 = n/d$.

(iii) Let $(m, n) = c$ and write c in the form $c = ma + nb$ where a and b are integers. Since $x^n = e$ for all $x \in G$, any solution of $x^m = e$ satisfies $x^c = x^{ma+nb} = e^a e^b = e$. Conversely, if $x^c = e$, then $x^m = e$, since $c \mid m$. Thus we need only count the solutions of $x^c = e$, and since $c \mid n$ there are clearly exactly c of them, namely, g^r for

$$r = \frac{n}{c}, \frac{2n}{c}, \ldots, \frac{cn}{c}.$$

(iv) Let $d \mid n$. Then the element $h = g^{n/d}$ has order exactly d and so generates a cyclic subgroup H of order d. We have to show that there are no others. Now every element of H satisfies $x^d = e$, and by (iii) the total number of solutions of $x^d = e$ in G is $(n, d) = d$. Hence all solutions are in H. In particular, every element of order d is in H and so all elements of order d generate the same subgroup H.

(v) This follows immediately. If $d \mid n$, all elements of order d are generators of the unique subgroup H of order d, so there are exactly $\phi(d)$ of them, by Theorem 7D. If $d \nmid n$, there are no elements of order d, by (i) or (ii).

COROLLARY. *For any positive integer n, $\sum_{d \mid n} \phi(d) = n$.*

Proof. The summation is over all distinct divisors d of n. By part (v) of the theorem, for each such d, $\phi(d)$ is the number of elements in G with order exactly d. Hence the sum is the total number of elements in G, since each element has order dividing n.

Example 7.8. Let $G = P_n$, the group of nth roots of 1 in $\mathbf{C}$. Then for each $d \mid n$, the subgroup of order d is P_d. Similarly, the solutions of $x^m = 1$ in G are all those mth roots of 1 which are also nth roots of 1. Thus they are the elements of $P_n \cap P_m$, which is a subgroup of each of P_n and P_m. By part (iii) of the theorem this group has order (m, n) so we must have $P_n \cap P_m = P_{(m,n)}$ for all m, n. One could prove this result directly in $\mathbf{C}$ and use it to give an alternative proof of the theorem which, on the face of it, seems simpler than the one given. Similarly one can "prove" the corollary by counting all the nth roots of 1 in $\mathbf{C}$ according to their orders: each one is a primitive dth root for a unique $d \mid n$, and there are $\phi(d)$ primitive dth roots in $\mathbf{C}$, all of which are amongst the nth roots. But this method is really cheating because it assumes the fact that all finite cyclic groups of all orders are subgroups of one big group T whose construction depends on the existence of the real or complex number system. This is a deeper and more complicated matter than the structure of cyclic groups. The natural way in which to consider all finite cyclic groups together is as *quotient* groups of one big group, namely $\mathbf{Z}$.

Exercises

1. Solve the following congruences: (i) $11x \equiv 18 \pmod{25}$; (ii) $36x \equiv 168 \pmod{192}$.

2. Find the smallest positive multiple of 84 whose last three digits (in decimal notation) are 832.

3. Solve the simultaneous congruences
$$\begin{cases} x \equiv 3 \pmod{18} \\ 3x \equiv 19 \pmod{35}. \end{cases}$$

4. Solve the simultaneous congruences
$$\begin{cases} 5x \equiv 21 \pmod{48} \\ 6x \equiv 10 \pmod{70} \\ 7x \equiv 35 \pmod{1001}. \end{cases}$$

5. Events A and B occur regularly at intervals of 4 days and 5 days, respectively. In a certain year, event A occurred on Sunday, 1st January and event B occurred on Monday, 2nd January. On how many occasions during the year did the two events occur together on a Thursday? When was the last such Thursday? (Answer in the form: on the nth day of the year.)

6. Prove that the simultaneous congruences
$$\begin{cases} x \equiv a \pmod{m} \\ x \equiv b \pmod{n} \end{cases}$$
have a solution if and only if (m, n) divides $a - b$.

7. Prove that an integer written in the scale of 10 is divisible by 9 if and only if the sum of its digits is divisible by 9. State and prove a similar rule for divisibility by 11.

8. Solve the simultaneous congruences
$$\begin{cases} 3x - 5y \equiv 7 \pmod{45} \\ 10x - 36y \equiv 12 \pmod{45}. \end{cases}$$

9. Prove that $\phi(n)$ is even for all $n > 2$.

10. Prove that $\phi(mn) \geq \phi(m) \phi(n)$ for all positive integers m and n.

11. Prove that if $n \geq 3$, the product of all the primitive nth roots of 1 in **C** is 1.

12. Prove that the primitive 12th roots of 1 in **C** are the roots of the equation $z^4 - z^2 + 1 = 0$. Find a corresponding equation for the primitive 10th roots of 1.

13. Prove that if G is a finite group of odd order then every element of G has a "square-root", i.e. $(\forall g \in G)(\exists h \in G)(h^2 = g)$. Does each element necessarily have a *unique* square-root? (Hint: think about cyclic subgroups of G.)

14. Prove that the congruence $x^2 \equiv 1 \pmod{n}$ has exactly two solutions modulo n if n is an odd prime number. Prove that it has exactly four solutions modulo n if $n = pq$, where p and q are distinct odd prime numbers.

15. Show that if n is a positive integer and $d \mid n$, then the number of

integers x such that $1 \leqslant x \leqslant n$ and $(x, n) = d$ is exactly

$$\phi\left(\frac{n}{d}\right).$$

16. Show that if a and b are integers and $ab \equiv 3 \pmod 4$ then either $a \equiv 3 \pmod 4$ or $b \equiv 3 \pmod 4$.

A sequence of integers $a_1, a_2, a_3, \ldots$ is defined by: $a_1 = 1$; $a_{n+1} = 4a_1 a_2 \ldots a_n + 3$. Prove that $(a_i, a_j) = 1$ if $i \neq j$ and deduce that there are infinitely many prime numbers of the form $4k + 3$.

CHAPTER 8

Rings and Fields

So far, we have confined our study of abstract algebraic systems to groups. In applications to the integers, for example, we have concentrated on the additive group structure and although we have been able to deduce facts about the multiplicative structure, such as the unique factorization theorem, this was only possible because multiplication of integers is definable in terms of addition. We cannot expect the same method to work in other circumstances; to make similar progress with polynomials, for example, or to prove deeper results about integers, we need to study the interplay between additive and multiplicative structures more closely. For this purpose, the basic notion is that of a ring.

Definition. A *ring* is set R with two binary operation, +, ·, and a unary operation, −, defined on it, and containing special elements 0 and 1, for which the following axioms hold:

(A1) $(x+y)+z = x+(y+z)$ for all $x, y, z \in R$;
(A2) $x+0 = 0+x = x$ for all $x \in R$;
(A3) $x+(-x) = (-x)+x = 0$ for all $x \in R$;
(A4) $x+y = y+x$ for all $x, y \in R$;
(M1) $(xy)z = x(yz)$ for all $x, y, z \in R$;
(M2) $x1 = 1x = x$ for all $x \in R$;
(AM1) $x(y+z) = xy + xz$ ⎫
   and $(x+y)z = xz + yz$ ⎭   for all $x, y, z \in R$.

These laws are a subset of the "standard laws" of Chapter 1 and can be summed up by saying that R is an additive Abelian group having an associative and distributive multiplication defined on it and containing an identity element 1 for multiplication.

We have already met several examples of rings. The most obvious are $\mathbf{Z, Q, R}$ and $\mathbf{C}$. Also, the algebra of all polynomials $p(X)$ with real coefficients is a ring (see Example 1.2). These rings all satisfy the extra axiom

(M4) $xy = yx$ for all $x, y \in R$

and are consequently called *commutative rings*. An example of a ring which is not commutative is the ring of all 2×2 matrices with real entries (see Example 1.3). More generally, the set of all $n \times n$ matrices with real entries is a non-commutative ring if $n \geqslant 2$.

We start with some elementary consequences of the axioms.

THEOREM 8A. *Let R be a ring. Then*

(i) $a \cdot 0 = 0 \cdot a = 0$ *for all $a \in R$;*

(ii) $(-a) \cdot b = a \cdot (-b) = -(a \cdot b)$ *for all $a, b \in R$;*

(iii) $(-a) \cdot (-b) = ab$ *for all $a, b \in R$;*

(iv)   *any product $a_1 a_2 \ldots a_n$ in R is independent of the bracketing of the factors;*

(v) $a^m a^n = a^n a^m = a^{m+n}$ *and* $(a^m)^n = a^{mn}$ *for all $a \in R$ and all $m, n \in \mathbf{N}$, where a^n is defined inductively by $a^0 = 1$, $a^{n+1} = a^n a$;*

(vi) $\left(\sum_{i=1}^{n} a_i\right) \cdot \left(\sum_{j=1}^{m} b_j\right) = \sum_{i=1}^{n} \sum_{j=1}^{m} (a_i b_j)$ *for all $a_i, b_j \in R$.*

Proof. (i) and (ii) were proved at the end of Chapter 1 using only the ring axioms. (iii) is an immediate consequence of (ii). The proof of (iv) is the same as that of Theorem 4A (vii). It is easy to deduce (v) from (iv) as in Theorem 4B. Finally (vi) is proved as follows. First use induction on n to show that $(a_1 + a_2 + \ldots + a_n)b = a_1 b + a_2 b + \ldots + a_n b$ for any $a_i, b \in R$. The inductive step uses one application of the distributive law (AM1). Now use this result, with $b = b_1 + b_2 + \ldots + b_m$, and apply induction on m, to finish the proof. We leave this as an exercise.

Two warnings are in order here. First, a^n is not defined in general for negative integers n; this is because we have not assumed that elements have inverses with respect to multiplication. Second, there is no cancellation law: $ab = ac$ does not imply $b = c$, even when $a \neq 0$. Indeed, it is possible for $ab = 0$ to be true when neither a nor b is 0. For example, in the ring of 2×2 matrices, the product of

$$\begin{pmatrix} 1 & 0 \\ 0 & 0 \end{pmatrix} \text{ and } \begin{pmatrix} 0 & 0 \\ 0 & 1 \end{pmatrix}$$

is the zero matrix.

Every ring contains certain special elements which in many ways play the role of integers. If x is any element of the ring R and if $n \in \mathbf{Z}$

then nx denotes, as usual, the nth additive power of x in the additive group of R. Thus $3x = x + x + x$ and $(-2)x = -x - x$. If R is the ring of integers itself then we have seen that nx can also be interpreted as the product of n and x in R. In general, we cannot do this, because $n \notin R$. However, if we write $\bar{n} = n1$, where 1 is the identity element of R, it is easy to prove that $\bar{n}x = nx$, where the left-hand side is now a product in R. For positive n this follows from the distributive law, since $(1 + 1 + \ldots + 1)x = x + x + \ldots + x$ (with n summands on each side). For $n = -m < 0$, one simply has to observe that

$$\bar{n} = (\overline{-m}) = (-m)1 = -(m1) = -\bar{m},$$

and hence

$$nx = (-m)x = -(mx) = -(\bar{m}x) = (-\bar{m})x = \bar{n}x.$$

In spite of the clear distinction between $n \in \mathbf{Z}$ and $\bar{n} \in R$, it is common practice to omit the bars in many contexts and to behave as if the integers were actually members of R. This causes no serious problems provided one remembers that it may happen that $\bar{m} = \bar{n}$ even when $m \neq n$. Thus, if bars are omitted, the equation $2 = 0$ may well be true in a ring. We shall return to this point later.

In a commutative ring various computations can be carried out which are not valid in arbitrary rings. A good example is the binomial theorem, which we will now prove in any commutative ring. This theorem is not true, for example, in the ring of matrices; for two $n \times n$ matrices A and B the distributive laws give

$$(A + B)^2 = A^2 + AB + BA + B^2,$$

but one cannot write $AB + BA = 2AB$ since $AB \neq BA$ in general.

THEOREM 8B. *Let R be a commutative ring. Then*

(i) $\quad (ab)^n = a^n b^n$ *for all* $a, b \in R$, $n \in \mathbf{N}$;

(ii) $\quad (a + b)^n = a^n + na^{n-1}b + \binom{n}{2}a^{n-2}b^2 + \ldots +$

$\binom{n}{r}a^{n-r}b^r + \ldots + b^n$, *for all $a, b \in R$, and all integers*

$n > 0$, *where* $\binom{n}{r}$ *denotes, as usual, the integer*

$n!/r!(n-r)!$.

Proof. (i) is proved exactly as for Abelian groups (but it is no longer valid for $n < 0$). (ii) is proved by induction on n in the same way as for real numbers. If $n = 1$ the result is trivial. We therefore suppose that the given formula is true for a fixed value of n (and for all $a, b \in R$) and deduce that

$$(a+b)^{n+1} = (a+b)^n(a+b)$$
$$= (a^n + na^{n-1}b + \ldots + \binom{n}{r}a^{n-r}b^r + \ldots + b^n)(a+b)$$
$$= (a^{n+1} + na^n b + \ldots + \binom{n}{r}a^{n-r+1}b^r + \ldots + ab^n)$$
$$+ (a^n b + na^{n-1}b^2 + \ldots + \binom{n}{r}a^{n-r}b^{r+1}$$
$$+ \ldots + b^{n+1}).$$

The terms in this expression are all of the form $ka^p b^q$ where k is an integer and $p + q = n + 1$. The coefficient of $a^{n+1-r}b^r$ is

$$\binom{n}{r} + \binom{n}{r-1}$$

where we interpret $\binom{n}{0}$ and $\binom{n}{n}$, as usual, to be 1, and $\binom{n}{n+1}$ to be 0.

An easy calculation shows that

$$\binom{n}{r} + \binom{n}{r-1} = \binom{n+1}{r},$$

and therefore

$$(a+b)^{n+1} = a^{n+1} + (n+1)a^n b + \ldots + \binom{n+1}{r}a^{n+1-r}b^r$$
$$+ \ldots + b^{n+1}.$$

This completes the inductive step and so proves the theorem.

We have already met the units of **Z**. In general, an element x of a ring R is called a *unit of R* if it has a multiplicative inverse, that is, if there exists $y \in R$ such that $xy = yx = 1$. The set of units is denoted by $U(R)$.

Example 8.1. $U(\mathbf{Z}) = \{\pm 1\}$.

Example 8.2. In each of the rings **Q, R, C**, all non-zero elements are units, so we have $U(\mathbf{Q}) = \mathbf{Q}^*$, $U(\mathbf{R}) = \mathbf{R}^*$, $U(\mathbf{C}) = \mathbf{C}^*$.

Example 8.3. In the ring $\mathbf{R}^{n \times n}$ of $n \times n$ real matrices, the units are the invertible (or non-singular) matrices. Thus $U(\mathbf{R}^{n \times n}) = \mathrm{GL}_n(\mathbf{R})$ (see Example 4.10).

The reader has probably noticed that in each of these examples the set of units is a familiar group. This is no accident, but an easy theorem.

THEOREM 8C. *The set $U(R)$ of units of any ring R is a group with respect to the ring multiplication.*

Proof. Write $U = U(R)$. If $x, y \in U$ then each has an inverse, say x', y', such that $xx' = x'x = yy' = y'y = 1$. It follows that the product xy in R also has an inverse, namely, $y'x'$; for

$$(xy)(y'x') = x(yy')x' = x1x' = xx' = 1$$

and similarly $(y'x')(xy) = 1$. Thus $xy \in U$ and multiplication induces a binary operation on U. This operation is associative because multiplication in R is associative. Now $1 \in U$ (its inverse is 1) and 1 acts as neutral element for multiplication. It remains to show that inverses exist in U. This is almost, but not quite, the definition of U. We know that any $x \in U$ has an inverse x' in R, and it will suffice to show that this x' is in U, i.e., that the inverse of a unit is a unit. But this is clear since x' has the inverse x in R.

Note. The most important property of units is that they can be cancelled: if $ua = ub$ in R, where u is a unit, then $a = b$, because we may multiply on the left by the inverse u^{-1} of u, giving $u^{-1}ua = u^{-1}ub$, whence $1a = 1b$, i.e., $a = b$. Similarly, $au = bu$ implies $a = b$ if u is a unit.

A *subring* of a ring R is defined in the obvious way by analogy with subgroups of a group. A subring is a subset S of R such that $0, 1 \in S$ and, if $a, b \in S$ then $a + b$, $-a$ and ab all lie in S. It follows that S is itself a ring with respect to the restricted operations.

Example 8.4. The set of all "integers" $\bar{n} = n1$ of R is a subring. The reader should check all details himself and do the same for all the following examples.

Example 8.5. In the chain of sets $\mathbf{Z} \subset \mathbf{Q} \subset \mathbf{R} \subset \mathbf{C}$, each is a subring of the next. However, $\mathbf{N}$ is not a subring of $\mathbf{Z}$.

Example 8.6. In the ring of 2×2 real matrices, the set of matrices of the form

$$\begin{pmatrix} a & 0 \\ b & c \end{pmatrix}$$

is a subring, but the set of matrices of the form

$$\begin{pmatrix} a & b \\ c & 0 \end{pmatrix}$$

is not a subring.

Example 8.7. The set of all even integers fails to be a subring of $\mathbf{Z}$ in only one respect: it does not contain the identity element 1. (Nor does it have an identity element of its own.) It should be mentioned here that some books do not require rings to have identity elements, and speak of rings-with-1 when they happen to possess them. If these alternative conventions are adopted then the set of even integers must be considered as a subring of $\mathbf{Z}$.

Notice that every subring of a ring R is, in particular an additive subgroup of R. Therefore, all theorems we have proved about subgroups of groups apply in this sense to subrings of rings. For example, in a finite ring with n elements, the number of elements in any subring is a divisor of n. Also, the additive cosets of a subring form a partition of the ring.

We shall continue to explore the analogy between groups and rings and we shall define such concepts as products of rings, isomorphisms and homomorphisms of rings, and quotient rings. There are one or two extra complications caused by the extra operation, but the arguments go through in a very similar way. The resulting theorems will be applied in the next two chapters to the ring of integers and to rings of polynomials.

If R and S are two rings then, since they are, in particular, Abelian groups with respect to addition, their set-theoretical product $R \times S$ is also an Abelian group with respect to the addition defined by $(r_1, s_1) + (r_2, s_2) = (r_1 + r_2, s_1 + s_2)$. If we now define multiplication in $R \times S$ by $(r_1, s_1)(r_2, s_2) = (r_1 r_2, s_1 s_2)$ then it is an easy matter to check that $R \times S$ is a ring, called the product of the rings R and S. Its zero element is $(0, 0)$ and its identity element is $(1, 1)$. We leave the verification as an exercise. If R and S are commutative rings then, clearly, so is $R \times S$.

THEOREM 8D. *Let R and S be rings. Then*

$$U(R \times S) = U(R) \times U(S).$$

Proof. The element $(r, s) \in R \times S$ is a unit if and only if there is an element $(r', s') \in R \times S$ such that $(r, s)(r', s') = (r', s')(r, s) = (1, 1)$. But $(r, s)(r', s') = (rr', ss')$ by definition; so (r, s) is a unit if and only if $\exists r' \in R$, $s' \in S$ such that $rr' = r'r = 1$ and $ss' = s's = 1$, that is, if and only if both r and s are units. This proves the theorem. Note that the equality $U(R \times S) = U(R) \times U(S)$ is not only equality of sets, but of groups: the multiplication in $U(R \times S)$ is the same as the multiplication in the product group $U(R) \times U(S)$.

Just as a homomorphism of groups is a map which preserves the group operations, so a *homomorphism of rings* is a map $f : R \to S$, where R and S are rings, which preserves the ring operations. Specifically, the conditions for f to be a homomorphism of rings are:

(i) $f(x + y) = f(x) + f(y)$ for all $x, y \in R$;

(ii) $f(xy) = f(x)f(y)$ for all $x, y \in R$;

(iii) $f(1) = 1$.

We have already seen that (i) implies $f(-x) = -f(x)$ and $f(0) = 0$, because R and S are additive groups (the reader should recall the argument, which was first given for isomorphisms of groups on p. 50). Unfortunately (ii) does not imply $f(1) = 1$, so we have to put this in as a separate assumption. (Example: the map from real numbers to matrices sending x to

$$\begin{pmatrix} x & 0 \\ 0 & 0 \end{pmatrix}$$

satisfies (i) and (ii), but not (iii).) However, (ii) and (iii) together do imply $f(x^{-1}) = f(x)^{-1}$ whenever x is a unit of R, because $f(x^{-1})f(x) = f(x^{-1}x) = f(1) = 1$, and similarly $f(x)f(x^{-1}) = 1$. This proves:

THEOREM 8E. *If $f : R \to S$ is a homomorphism of rings then f sends units of R to units of S and preserves inverses. Hence f induces a homomorphism of multiplicative groups from $U(R)$ to $U(S)$.*

An *isomorphism of rings* is, of course, a bijective homomorphism of rings. Two rings which are isomorphic therefore have essentially the same ring structure.

Example 8.8. If R is any ring the map $f : \mathbf{Z} \to R$ given by $f(n) = n1 = \bar{n}$ is a homomorphism of rings. If R is, for example, the ring of real numbers or the ring of 2×2 real matrices, this homomorphism is an injection but not a surjection.

Example 8.9. The map $f : \mathbf{C} \to \mathbf{R}^{2 \times 2}$ defined by

$$f(x + iy) = \begin{pmatrix} x & y \\ -y & x \end{pmatrix}$$

for all real x and y is a homomorphism of rings and is an injection. The reader should check all the details of this example.

We now come to the construction of quotient rings, and we follow the same plan as for groups. We are looking for a partition of a ring R in which the classes themselves can be added *and multiplied* to give a ring structure on the quotient set. We already know that in order to make addition work correctly we must start by taking the cosets of an additive subgroup A of R. Here we simply use the fact that R is an additive *Abelian* group. Theorem 6D shows that for any additive subgroup A of R the set R/A of all additive cosets $\langle x \rangle = A + x$ is itself an Abelian group with respect to an addition defined by $\langle x \rangle + \langle y \rangle = \langle x + y \rangle$. However, if we now try to define multiplication by $\langle x \rangle \langle y \rangle = \langle xy \rangle$ there is a slight difficulty, because the right-hand side will not necessarily be independent of the representatives chosen for the classes $\langle x \rangle$ and $\langle y \rangle$. We need to satisfy the condition of Theorem 6C:

$$\langle x \rangle = \langle x' \rangle \ \& \ \langle y \rangle = \langle y' \rangle \Rightarrow \langle xy \rangle = \langle x'y' \rangle,$$

which is not true for all subgroups. (For example if $R = \mathbf{Q}$ and $A = \mathbf{Z}$, then $\langle 1 \rangle = \langle 2 \rangle$ and $\langle \tfrac{1}{2} \rangle = \langle -\tfrac{1}{2} \rangle$, but $\langle 1 \cdot \tfrac{1}{2} \rangle \ne \langle 2 \cdot (-\tfrac{1}{2}) \rangle$.) A special case of this consistency condition says that if $\langle x \rangle = \langle 0 \rangle$ then, for any y, $\langle xy \rangle = \langle 0y \rangle = \langle 0 \rangle$; in other words, if $x \in A$ then, for any $y \in R$, $xy \in A$, and similarly $yx \in A$. Any additive subgroup A which satisfies this last condition (viz. $x \in A \ \& \ y \in R \Rightarrow xy \in A \ \& \ yx \in A$) is called an *ideal* of R. We shall see that this condition is in fact all that is needed to make quotients work smoothly. Note that an ideal is closed under multiplication, but is not always a subring because we do not require that $1 \in A$. Indeed, if A is an ideal and $1 \in A$ then we must have $A = R$.

So let us now suppose that A is any ideal of R, and suppose that $\langle x \rangle = \langle x' \rangle$ and $\langle y \rangle = \langle y' \rangle$. This means that $x - x' \in A$ and $y - y' \in A$, and we would like to know that $\langle xy \rangle = \langle x'y' \rangle$, that is, $xy - x'y' \in A$. Now we can write $xy - x'y' = x(y - y') + (x - x')y'$ in any ring. Since $y - y' \in A$ and A is an *ideal*, we have $x(y - y') \in A$. Similarly $(x - x')y' \in A$ and it follows that $xy - x'y' \in A$ i.e. $\langle xy \rangle = \langle x'y' \rangle$. This shows that multiplication can be defined on R/A by the rule $\langle x \rangle \langle y \rangle = \langle xy \rangle$ without ambiguity. We can now deduce very easily that R/A is a ring with identity element $\langle 1 \rangle = A + 1$. Clearly

$$\langle 1 \rangle \langle x \rangle = \langle x \rangle \langle 1 \rangle = \langle x \rangle \quad \text{for all } x \in R,$$

and

$$(\langle x \rangle \langle y \rangle)\langle z \rangle = \langle xy \rangle \langle z \rangle = \langle (xy)z \rangle = \langle x(yz) \rangle = \langle x \rangle \langle yz \rangle = \langle x \rangle (\langle y \rangle \langle z \rangle).$$

The distributive laws are proved by a similar routine argument. Also, if $xy = yx$ in R, then $\langle x \rangle \langle y \rangle = \langle y \rangle \langle x \rangle$ in R/A. This proves:

THEOREM 8F. *If A is any ideal of a ring R then the set R/A of additive cosets of A in R is a ring with respect to operations defined by*

$$\langle x \rangle + \langle y \rangle = \langle x + y \rangle$$
$$\langle x \rangle \cdot \langle y \rangle = \langle xy \rangle$$

where $\langle x \rangle$ denotes the coset $A + x$. The zero element of R/A is $\langle 0 \rangle = A$ and its identity element is $\langle 1 \rangle = A + 1$. If R is a commutative ring then so is R/A.

As in the case of groups, the quotient map $q : R \to R/A$ is clearly a homomorphism, this time of *rings*; this is simply a reflection of the way we defined the ring operations in R/A. We can now more or less copy the First Isomorphism Theorem for groups (Theorem 6G) with appropriate changes to ring terminology. The *kernel* of a homomorphism of rings $f : R \to S$ is the set of elements $x \in R$ such that $f(x) = 0$. The *image* of f is just its set-theoretical image $f(R)$.

THEOREM 8G. *Let R and S be rings and let $f : R \to S$ be a homomorphism of rings. Then*

(i) *the kernel K of f is an ideal of R;*

(ii) *the fibres of f are the additive cosets of K;*

(iii) *the image $T = f(R)$ of f is a subring of S;*

(iv) *$T \cong R/K$ (an isomorphism of rings).*

Proof. Because f is a homomorphism of additive groups, K is an additive subgroup of R and the fibres of f are the additive cosets of K. Also T is an additive subgroup of S, and $T \cong R/K$ *as groups*. All this is contained in Theorem 6G. Now if $k \in K$ and $r \in R$, then $f(k) = 0$, so $f(kr) = f(k)f(r) = 0 \cdot f(r) = 0$ and similarly $f(rk) = 0$; hence $kr \in K$ and $rk \in K$, so K is an *ideal* of R. Similarly, if $x, y \in T$, then $x = f(a)$ and $y = f(b)$ for some $a, b \in R$, so $xy = f(a)f(b) = f(ab) \in T$. Also $1 = f(1) \in T$, and this shows that T is a subring of S. Finally, the group isomorphism $f^* : R/K \to T$ which, we recall, is defined by $f^*(\langle r \rangle) = f(r)$, is in fact a ring isomorphism because
$f^*(\langle r_1 \rangle \langle r_2 \rangle) = f^*(\langle r_1 r_2 \rangle) = f(r_1 r_2) = f(r_1)f(r_2) = f^*(\langle r_1 \rangle)f^*(\langle r_2 \rangle)$ and, of course, $f^*(\langle 1 \rangle) = f(1) = 1$. This completes the proof.

Before looking at the applications of these results we shall introduce some new concepts in ring theory which have no counterparts in group theory. They are concerned with the construction of fractions and are important for a proper understanding of rational numbers and rational functions.

We start with the notion of zero-divisor in a *commutative* ring R. Obviously, every element x of R divides zero in the sense that $xy = 0$ for a suitable y (namely, $y = 0$). To avoid this triviality we therefore define a *zero-divisor* in R to be an element $x \in R$ satisfying: $\exists y \in R$ such that $xy = 0$ and $y \neq 0$. (In a non-commutative ring we would have to distinguish between left and right zero-divisors, but we shall not consider this case.)

An element x which is *not* a zero divisor satisfies the condition: $xy = 0 \Rightarrow y = 0$. Such an element x can be cancelled from equations of the form $xa = xb$, because $xa = xb \Rightarrow x(a - b) = 0 \Rightarrow a - b = 0 \Rightarrow a = b$. On the other hand, a zero-divisor x definitely cannot always be cancelled, because there is a $y \neq 0$ such that $xy = 0 = x0$. Thus any commutative ring is partitioned into two types of elements: the zero-divisors and the *cancellable* elements. Amongst the zero-divisors we distinguish the *proper zero-divisors*, namely those which are not equal to 0.

THEOREM 8H. *Let R be a commutative ring. Then the following three conditions are equivalent:*

(i) R *contains no proper zero-divisors*;

(ii) $xy = 0$ *in R implies $x = 0$ or $y = 0$*;

(iii) R *satisfies the cancellation law:*— *if $xa = xb$ in R and $x \neq 0$ then $a = b$.*

Proof. (ii) is simply a re-wording of (i) in a more symmetric form, and is obviously equivalent to it. Also, we showed above than an element is cancellable if and only if it is not a zero-divisor. Thus (i) says precisely that all non-zero elements are cancellable, which is the statement (iii).

Definition. An *integral domain* is a commutative ring which satisfies one (and therefore all) of the conditions of Theorem 8H and in which the elements 0 and 1 are distinct. (We have not previously assumed $1 \neq 0$, and it is easy to verify that there is, up to isomorphism, just one ring in which $1 = 0$. This ring has only one element. It is a moot point whether this ring satisfies the conditions of the theorem, but a careful logical analysis should convince the reader that it does. We now wish to exclude this trivial ring when considering integral domains, although we must allow that it really is a ring. If we imposed $1 \neq 0$ as an axiom for rings in general then Theorem 8G would be false, because the whole ring R is certainly an ideal of R and the corresponding quotient R/R has only one element!) Returning to our definition, an integral domain is a commutative ring, with at least two elements, in which the cancellation law holds. Examples that spring to mind are **Z**, **Q**, **R** and **C**. We shall see in Chapter 10 that certain rings of polynomials are also integral domains.

The reason that **Q**, **R** and **C** are integral domains is that in each of them, any non-zero element is a unit. We have already shown that units are always cancellable (see the Note on p. 99) so the cancellation law follows immediately. Such rings are worth a special name and are called *fields*. A field is thus a ring R which satisfies, in addition to the ring axioms laid down at the beginning of the chapter, the following laws:

(M3)   every $x \neq 0$ in R has an inverse $x^{-1} \in R$ such that $xx^{-1} = x^{-1}x = 1$;

(M4) $xy = yx$ for all $x, y \in R$;

(AM2) $1 \neq 0$.

This brings us full circle, because these three laws, together with the ring axioms, comprise the complete set of laws of "standard algebra" listed in Chapter 1. If the whole list is now studied again it will be seen that a field can be described very briefly in terms of two group structures, additive and multiplicative. To be precise, we may restate the definition of a field as follows. A field F is a set with two binary operations + and · defined on it such that:

(i) F is an Abelian group with respect to addition;

(ii)   if the zero element of this Abelian group is removed, the remaining elements form an Abelian group with respect to multiplication;

(iii)   multiplication is distributive over addition.

Note that these three conditions imply $1 \neq 0$ because the multiplicative group specified in (ii) does not contain 0, but does contain 1.

After groups, fields are probably the most important of the many types of abstract algebra which are studied today. They are the domains in which one can apply all the operations and rules of standard algebra. They appear frequently as the source of coefficients for polynomial and linear equations, and as the proper place to seek solutions of such equations. They occur in number theory as nicely structured sets of algebraic numbers and in geometry as sets of possible coordinates of points. In all these contexts it is their richness of structure and ease of algebraic manipulation that make them useful.

All non-zero elements of a field are cancellable, and this property remains true in any subring. Thus all subrings of fields are integral domains. One must beware of applying arguments like this unthinkingly. For example, all non-zero elements of a field are units, but this property does not remain true in a subring because a subring may contain a given element without containing its inverse. The simplest example is the subring **Z** of the field **Q**; it is an integral domain (in fact the prototype!) but it has only two units 1 and -1.

106   RINGS AND FIELDS

Example 8.10. In the field **C**, the set of all Gaussian integers $m + in$, where $m, n \in \mathbf{Z}$, is a subring and is therefore an integral domain.

Example 8.11. In the field **R**, the set S of all numbers of the form $a + b\sqrt{2}$, where a and b are *rational* numbers, is a subring because 1 and 0 are of this form and so are the sum and product of any two numbers of the given form. (Note that
$(a + b\sqrt{2})(a' + b'\sqrt{2}) = (aa' + 2bb') + (ab' + ba')\sqrt{2}$, and the numbers $aa' + 2bb'$, $ab' + ba'$ are rational if a, a', b, b' are rational.) This subring S of **R** is actually a *subfield*. The reason for this is that if a and b are rational and not both zero then the inverse of $a + b\sqrt{2}$ in **R** can be written in the form

$$\frac{a - b\sqrt{2}}{a^2 - 2b^2} = \frac{a}{a^2 - 2b^2} + \frac{-b}{a^2 - 2b^2}\sqrt{2}$$

and so is actually a member of S. Thus every non-zero element of S is a unit in S, and S is a field.

The relationship between integral domains and fields is even closer than we have suggested so far. Not only is every subring of a field an integral domain, but every integral domain is actually a subring of a suitable field. We end the chapter by proving this important result. Given an integral domain D we have to construct a field F such that D is contained in F as a subring. In fact we shall do something less precise, namely, construct a field F such that D is *isomorphic* to a subring of F, i.e., such that there is an injective homomorphism $D \to F$. However, having done this, we can modify F (somewhat artificially) so that it actually contains D as a subring; we simply remove from F the isomorphic copy D' of D and replace it by D. The new set F' can then be made into a field by a patching-up process in an obvious way so that D is a subring. This process is called "identifying D with its image in F" and we shall not write out its formal details.

THEOREM 8J. *Let D be an integral domain. Then there is a field F containing D as a subring such that every element of F is of the form xy^{-1} where $x, y \in D$. The field F is unique up to isomorphism and has the following universal property: if $\theta : D \to F'$ is any embedding of D in a field (i.e., any injective ring homomorphism into a field) then θ can be extended uniquely to an embedding of F in F'.*

Proof.   The clue to the construction of F is the form xy^{-1} of its elements. We shall define F as a set of "fractions" x/y, where $x, y \in D$, but some care is needed in defining these. The familiar rules for manipulating fractions tell us that the denominator must never be zero and that x/y and x'/y' are the same fraction if $xy' = yx'$. We therefore use our experience of equivalence relations to define fractions as equivalence classes of pairs (x,y). The details are as follows. Let $S \subset D \times D$ be the set of all pairs (x,y), where $x, y \in D$ and $y \neq 0$. On S, define a relation $\sim$ by the rule

$$(x, y) \sim (x', y') \text{ if } xy' = yx' \text{ in } D.$$

We first verify that $\sim$ is an equivalence relation. Clearly it is reflexive since $xy = yx$. Also, it is symmetric, because if $xy' = yx'$ then $x'y = y'x$, again using the commutative law in D. The transitive law is a little harder. Suppose that $(x, y) \sim (x', y') \sim (x'', y'')$ in S. Then $xy' = yx'$ and $x'y'' = y'x''$, whence $xy'y'' = yx'y'' = yy'x''$. Since D is an integral domain and since, by definition, $y' \neq 0$, we may cancel y' to obtain $xy'' = yx''$, i.e., $(x, y) \sim (x'', y'')$. The equivalence classes of $\sim$ on S are called *fractions* and we write x/y for the equivalence class containing (x, y). We let F be the set of all these fractions, i.e., $F = S/\sim$, and we have to make F into a field by defining appropriate operations on fractions. First addition: we define

$$\frac{x}{y} + \frac{x'}{y'} = \frac{xy' + yx'}{yy'}. \tag{1}$$

Once again we are faced with the question whether this operation is well-defined, since it is an operation on a quotient set. We refer the reader to Theorem 6C and we make the necessary verification. Suppose that $(x_1, y_1) \sim (x_2, y_2)$ and $(x_1', y_1') \sim (x_2', y_2')$ in S. We have to show that $(x_1 y_1' + y_1 x_1', y_1 y_1') \sim (x_2 y_2' + y_2 x_2', y_2 y_2')$ in S. Now D is a ring, so both these pairs lie in $D \times D$; also since D is an integral domain and y_1, y_2, y_1' and y_2' are all non-zero, we have $y_1 y_1' \neq 0$ and $y_2 y_2' \neq 0$, so the pairs are in S. It remains to show that $(x_1 y_1' + y_1 x_1')(y_2 y_2') = y_1 y_1'(x_2 y_2' + y_2 x_2')$ in D. This follows from the given relations $x_1 y_2 = y_1 x_2$ and $x_1' y_2' = y_1' x_2'$ because

$$(x_1 y_1' + y_1 x_1')(y_2 y_2') = x_1 y_2 y_1' y_2' + y_1 y_2 x_1' y_2'$$
$$= y_1 x_2 y_1' y_2' + y_1 y_2 y_1' x_2'$$
$$= y_1 y_1'(x_2 y_2' + y_2 x_2').$$

Having established that addition is well-defined on F by equation (1), we verify that it makes F an Abelian group. Clearly $0/1$ acts as zero element and the addition is commutative. The associative law is proved by checking that the two ways of bracketing

$$\frac{x}{y} + \frac{x'}{y'} + \frac{x''}{y''}$$

both lead to the fraction

$$\frac{xy'y'' + yx'y'' + yy'x''}{yy'y''}.$$

As for additive inverses, one sees that

$$\frac{x}{y} + \frac{-x}{y} = \frac{0}{yy'} \quad \text{and} \quad \frac{0}{yy'} = \frac{0}{1}$$

since $0 \cdot 1 = yy' \cdot 0$. Thus $-x/y$ is inverse to x/y.

We define multiplication in F by the rule

$$\frac{x}{y} \cdot \frac{x'}{y'} = \frac{xx'}{yy'} \tag{2}$$

for all $x, x', y, y' \in D$ with $y, y' \neq 0$. We leave the reader to verify that (2) is unambiguous and that it defines an operation on F. It is clearly commutative and associative and has neutral element $1/1$. It is easy to check the distributive law:

$$\left(\frac{x}{y} + \frac{x'}{y'}\right)\frac{x''}{y''} = \frac{(xy' + yx')x''}{yy'y''}$$

$$= \frac{(xy' + yx')x''y''}{yy'y''y''}$$

$$= \frac{xx''}{yy''} + \frac{x'x''}{y'y''}.$$

Thus F is a commutative ring and it remains to show that each non-zero element x/y is a unit. But if $(x/y) \neq (0/1)$ then $x \neq 0$, so $(y/x) \in F$ and clearly

$$\frac{x}{y} \cdot \frac{y}{x} = \frac{xy}{xy} = \frac{1}{1}.$$

This shows that *F is a field*.

Now F does not contain D, but there is an obvious map $\sigma : x \mapsto x/1$ from D to F which is an injective homomorphism of rings. (Check this!) We therefore identify D with its image in F as described above, writing x instead of $x/1$ from now on. In this notation, for any $y \neq 0$ in D,

$$y^{-1} = \left(\frac{y}{1}\right)^{-1} = \frac{1}{y},$$

and hence

$$\frac{x}{y} = xy^{-1}$$

as claimed.

Next we prove the universal property of F. If $\theta : D \to F'$ is an embedding of D in a field, we see that, for $y \neq 0$ in D, $\theta(y) \neq 0$, so $\theta(y)$ has an inverse in F'. If θ is to be extended to an embedding $\theta^* : F \to F'$,

θ^* must preserve inverses, so $\theta^*(y^{-1}) = \theta(y)^{-1}$ and we must have
$$\theta^*(xy^{-1}) = \theta(x)\theta(y)^{-1} \qquad (3)$$
This shows that θ^*, if it exists, is unique. Again, to prove existence, we must show that (3) is unambiguous, i.e., if
$$\frac{x}{y} = \frac{x'}{y'}$$
in F, we must show that $\theta(x)\theta(y)^{-1} = \theta(x')\theta(y')^{-1}$ in F'. This we leave to the reader, who should also check that the map θ^* defined by (3) does in fact preserve addition and multiplication and send 1 to 1. The fact that θ^* is an *injection* follows from the fact that if $\theta^*(xy^{-1}) = 0$ then $\theta(x)\theta(y)^{-1} = 0$ in F' and so $\theta(x) = 0$. But this implies $x = 0$, since θ is an injection. Hence Ker $\theta^* = 0$ and θ^* is an injection by Theorem 6G, Corollary. The uniqueness of F now follows easily; if D is embedded in a field F' by a map θ, in such a way that every element of F' can be written $\theta(x)\theta(y)^{-1}$ for suitable $x, y \in D$, then the map $\theta^* : F \to F'$ is surjective and is therefore an isomorphism.

The field F constructed from D is called the *field of fractions of D*. It is of great importance and, in spite of the lengthy proof of its existence, it is essentially a simple and obvious construction. The two main examples (rational numbers and rational functions) will be discussed in detail in the next two chapters.

Exercises

1. Prove that a ring R is commutative if and only if the equation $x^2 - y^2 = (x - y)(x + y)$ is true for all $x, y \in R$.

2. Prove that the group of units of the ring of Gaussian integers (Example 8.10) is a cyclic group of order 4.

3. Prove that the set of all real numbers of the form $m + n\sqrt{2}$, where $m, n \in \mathbf{Z}$, is an integral domain. Show that $m + n\sqrt{2}$ is a unit in this ring if and only if $m^2 - 2n^2 = \pm 1$.

4. Formulate and prove for rings the analogue of Theorem 6H.

5. Prove that the only ideals of a field F are $\{0\}$ and F itself. Hence show that every ring homomorphism between two fields is an injection.

6. Prove that the ring M of all $n \times n$ real matrices has only two ideals $\{0\}$ and M itself. (Hint: consider the effect of multiplying members of an ideal by (i) the basic matrices E_{ij} with 0's in all places but one and (ii) the scalar matrices λI.)

7. Show that if D is an integral domain and $d \in D$ is a fixed non-zero element, then the map $x \mapsto dx$ from D to D is an injection. For

finite D deduce that this map is a bijection and hence prove that every finite integral domain is a field.

8. Let R be an arbitrary ring and let $\sigma : \mathbf{Z} \to R$ be the canonical homomorphism given by $\sigma(n) = \bar{n} = n1$. By considering the kernel of σ, show that there is a unique integer $k \geq 0$ (called the characteristic of R) such that (i) $kx = 0$ for every $x \in R$, and (ii) if $mx = 0$ for every $x \in R$ then $k \mid m$. Prove that if R is an integral domain then its characteristic is 0 or a prime number. Deduce that all non-zero elements of an integral domain have the same additive order.

9. Prove that if R and S are two commutative rings with at least two elements each, then $R \times S$ cannot be an integral domain.

CHAPTER 9

The Rings $\mathbf{Z_n}$ and the Field $\mathbf{Q}$

The simplest ring at our disposal is the ring $\mathbf{Z}$ of integers, and we now examine the rings which can be obtained from it by applying the abstract constructions of Chapter 8.

First, the subrings of $\mathbf{Z}$ are easy to determine. Every subring contains 1 and so must contain the additive subgroup generated by 1, which is $\mathbf{Z}$ itself. Thus $\mathbf{Z}$ is the only subring.

Next, the ideals of $\mathbf{Z}$ are also familiar already. Any ideal is an additive subgroup, so must be of the form $n\mathbf{Z}$ for some integer $n \geq 0$ (Theorem 4C). Conversely, each subgroup $n\mathbf{Z}$ is an ideal of $\mathbf{Z}$; for if $m \in n\mathbf{Z}$ then $n \mid m$, so $n \mid mr$ for all integers r, whence $mr \in n\mathbf{Z}$. This last argument works in any commutative ring R: if $a \in R$ then the set $aR = \{ar; r \in R\}$ is an ideal of R. Ideals formed in this way are called *principal ideals* and the element a is called a *generator* of the ideal aR. Theorem 4C therefore asserts that each ideal of $\mathbf{Z}$ is a principal ideal $n\mathbf{Z}$ generated by some $n \geq 0$.

Now, with respect to each of these ideals $n\mathbf{Z}$, for $n \geq 1$, we can form a quotient ring $\mathbf{Z}/n\mathbf{Z}$, which is denoted by $\mathbf{Z}_n$. (We exclude from consideration the case $n = 0$ since $\mathbf{Z}/0\mathbf{Z} = \mathbf{Z}/\{0\} \cong \mathbf{Z}$ is not a new ring). We have already met $\mathbf{Z}_n$ as an additive group. It is the group of residue classes modulo n and it is a cyclic group of order n generated by the residue class $\langle 1 \rangle$ containing 1. The new fact supplied by Theorem 8F is that, because $n\mathbf{Z}$ is an ideal, *multiplication* of residue classes is also possible according to the rule $\langle x \rangle \langle y \rangle = \langle xy \rangle$, and the group of residue classes thereby becomes a ring. This ring $\mathbf{Z}_n$ is commutative because $\mathbf{Z}$ is commutative, its identity element is $\langle 1 \rangle$ and its "integers" are $\bar{r} = r\langle 1 \rangle = \langle r \rangle$. We shall therefore write $\bar{r}$ from now on for the residue class $\langle r \rangle$ in $\mathbf{Z}_n$. Note that $\bar{r}$ is the image of r under the quotient map $\mathbf{Z} \rightarrow \mathbf{Z}_n$ whose kernel is $n\mathbf{Z}$.

111

Example 9.1. The ring $\mathbf{Z}_4$ has addition and multiplication tables as follows.

+	$\bar{0}$	$\bar{1}$	$\bar{2}$	$\bar{3}$
$\bar{0}$	$\bar{0}$	$\bar{1}$	$\bar{2}$	$\bar{3}$
$\bar{1}$	$\bar{1}$	$\bar{2}$	$\bar{3}$	$\bar{0}$
$\bar{2}$	$\bar{2}$	$\bar{3}$	$\bar{0}$	$\bar{1}$
$\bar{3}$	$\bar{3}$	$\bar{0}$	$\bar{1}$	$\bar{2}$

×	$\bar{0}$	$\bar{1}$	$\bar{2}$	$\bar{3}$
$\bar{0}$	$\bar{0}$	$\bar{0}$	$\bar{0}$	$\bar{0}$
$\bar{1}$	$\bar{0}$	$\bar{1}$	$\bar{2}$	$\bar{3}$
$\bar{2}$	$\bar{0}$	$\bar{2}$	$\bar{0}$	$\bar{2}$
$\bar{3}$	$\bar{0}$	$\bar{3}$	$\bar{2}$	$\bar{1}$

It would be tedious in the extreme to verify from these tables that the ring axioms are satisfied; but this is unnecessary since our analysis of quotient rings guarantees it in advance. The operations are performed by first applying them in $\mathbf{Z}$ and then reducing modulo 4. The same applies, of course, for any n. Note that $\mathbf{Z}_4$ is not an integral domain since $\bar{2} \cdot \bar{2} = \bar{0}$. Its units are the elements $\bar{1}$ and $\bar{3}$, which form a cyclic group of order 2.

Obviously, no two of the rings $\mathbf{Z}_n(n = 1, 2, 3, \ldots,)$ are isomorphic since no two have the same number of elements. We therefore have constructed infinitely many genuinely different finite rings. Computation in these rings is often referred to as "modular arithmetic" and can legitimately use all the laws of commutative rings, but not the law M3 and not the cancellation law except in special cases. To determine which of the rings $\mathbf{Z}_n$ obey these extra laws we must find the units of $\mathbf{Z}_n$.

THEOREM 9A. *The element $\bar{r}$ of the ring $\mathbf{Z}_n$ is a unit if and only if r is coprime to n.*

Proof. Since $\mathbf{Z}_n$ is a commutative ring,

$\bar{r}$ is a unit of $\mathbf{Z}_n \Leftrightarrow \bar{r} \cdot \bar{s} = \bar{1}$ in $\mathbf{Z}_n$ for some $s \in \mathbf{Z}$

$\Leftrightarrow \overline{rs} = \bar{1}$ in $\mathbf{Z}_n$ for some $s \in \mathbf{Z}$

$\Leftrightarrow rs \equiv 1 (\bmod n)$ for some $s \in \mathbf{Z}$

$\Leftrightarrow (r, n) = 1$, by Theorem 8B.

COROLLARY. $\mathbf{Z}_n$ *is a field if and only if n is a prime number.*

Proof. If n is a prime number then $(r, n) = 1$ for every r not divisible by n; hence in $\mathbf{Z}_n$, every $\bar{r} \neq \bar{0}$ is a unit, so $\mathbf{Z}_n$ is a field. On the other hand, if $n > 1$ is not prime, then $n = r_1 r_2$ for suitable integers r_1, r_2 satisfying $1 < r_1 < n$ and $1 < r_2 < n$. In this case $\bar{r}_1 \bar{r}_2 = \bar{n} = \bar{0}$ in $\mathbf{Z}_n$, but $\bar{r}_1 \neq \bar{0}$ and $\bar{r}_2 \neq \bar{0}$, so $\mathbf{Z}_n$ is not even an integral domain, let alone a field.

We have now constructed an infinite collection of finite fields Z_2, Z_3, Z_5, Z_7, Z_{11}, ... (the collection is infinite by Exercise 13 of Chapter 5). Arithmetic modulo a *prime* number obeys all the laws of standard algebra, and this explains the remarks in Chapter 7 about the solution of simultaneous congruences in several unknowns relative to a single modulus. If the modulus happens to be prime then all the usual tricks of elimination are available and the solutions can be found by standard algorithms for linear equations. We remark in passing that the fields Z_p for prime numbers p are not the only finite fields. There is in fact one with q elements whenever q is a power of a prime number, but it is rather harder to construct when q is not prime.

Example 9.2. The elements $0, 1, a, b$ with addition and multiplication defined by the tables

+	0	1	a	b
0	0	1	a	b
1	1	0	b	a
a	a	b	0	1
b	b	a	1	0

×	0	1	a	b
0	0	0	0	0
1	0	1	a	b
a	0	a	b	1
b	0	b	1	a

form a field. A method of establishing this fact without directly verifying all the laws will be given in the next chapter.

We denote by U_n the multiplicative group of units Z_n. By Theorem 9A the elements of U_n are the residue classes $\bar{r}$, where $1 \leq r \leq n$ and $(r, n) = 1$. Hence the number of elements in U_n is $\phi(n)$, where ϕ is Euler's function (see p. 89). (Note that the units of Z_n are the same things as the generating elements of the additive group of Z_n. See Exercise 12 at the end of this chapter for the analogous statement for general rings.) Now in a finite group of order m, every element has order dividing m and therefore its mth power is the identity (Theorem 4F, Corollary 2). If we apply this to the group U_n of order $\phi(n)$ we see that in Z_n, $\bar{r}^{\phi(n)} = \bar{1}$ for every *unit* $\bar{r}$. Translating this statement into the language of congruences, we obtain

THEOREM 9B. *(Euler's Theorem). Let n be a positive integer. Then*

$$r^{\phi(n)} \equiv 1 \pmod{n}$$

for all integers r coprime to n.

Proof. If $(r, n) = 1$ then $\bar{r}$ is a unit in Z_n, by Theorem 9A. Hence by the argument above, $\bar{r}^{\phi(n)} = \bar{1}$ in Z_n, that is, $r^{\phi(n)} \equiv 1 \pmod{n}$.

COROLLARY 1. *(Fermat's Theorem). Let p be a prime number. Then*

$$r^{p-1} \equiv 1 \pmod{p}$$

for all integers r not divisible by p.

Proof. When p is prime, $\phi(p) = p - 1$. Also, when p is prime, $(r, p) = 1 \Leftrightarrow p \nmid r$.

COROLLARY 2. *(Alternative form of Fermat's Theorem.) Let p be a prime number. Then $n^p \equiv n \pmod{p}$ for all integers n.*

Proof. An exercise.

These two results contain a wealth of arithmetic information, for instance, the by no means obvious facts that $5^6 - 1$ is divisible by 7 and that $17^4 - 1$ is divisible by 12. The proof we have given is the one which gives most insight into the reasons why such congruences are true. The results were originally proved by quite different methods, and it is perhaps worth giving a more elementary proof of Fermat's theorem. Let p be prime and let r be any integer not divisible by p. Then the integers $r, 2r, 3r, \ldots, (p-1)r$ lie in distinct residue classes modulo p (because if $rx \equiv ry \pmod{p}$ then $x \equiv y \pmod{p}$, r being coprime to p). Hence these numbers lie one in each of the non-zero residue classes and so their product is in the same residue class as $1 \cdot 2 \cdot 3 \ldots (p-1)$. In other words,

$$r^{p-1} \cdot 1 \cdot 2 \cdot 3 \ldots (p-1) \equiv 1 \cdot 2 \cdot 3 \ldots (p-1) \pmod{p}$$

and we deduce that $r^{p-1} \equiv 1 \pmod{p}$ because we can cancel each of $1, 2, 3, \ldots, p-1$. (The thoughtful reader will recognise that this proof can essentially be carried out in any Abelian group G : multiplication by $r \in G$ *permutes* the elements of G, so does not alter their product, and this proves Theorem 4F, Corollary 2 in the Abelian case. So the two proofs are not so different after all!)

Example 9.3. One can deduce many further congruences from Fermat's and Euler's theorems. For example, $n^{13} \equiv n \pmod{2{,}730}$ for all integers n. To prove this, observe that $2{,}730 = 2 \cdot 3 \cdot 5 \cdot 7 \cdot 13$ is a product of distinct primes, so it will be enough to show that $n^{13} \equiv n \pmod{p}$ for $p = 2, 3, 5, 7$ and 13 and for all n. If $p = 13$, Fermat's theorem tells us that $n^{13} \equiv n \pmod{p}$. For $p = 7$, we have $n^7 \equiv n \pmod 7$, whence $n^{13} = n^7 \cdot n^6 \equiv n \cdot n^6 \equiv n^7 \equiv n \pmod 7$. For $p = 5$, we argue: $n^{13} = n^5 \cdot n^8 \equiv n \cdot n^8 = n^5 \cdot n^4 \equiv n \cdot n^4 = n^5 \equiv n \pmod 5$ and similar arguments give the general congruences $n^{13} \equiv n \pmod p$ for $p = 2$ and 3.

From now on, we shall omit the bars on elements of $\mathbf{Z}_n$ except where they are necessary for clarity. Thus, for example, we shall write $r = s$ in $\mathbf{Z}_n$, meaning $r \equiv s \pmod{n}$, and say that 6 is a unit in $\mathbf{Z}_{13}$, with inverse $11 = -2$.

The units of $\mathbf{Z}_n$ can be described alternatively as the roots of 1 in $\mathbf{Z}_n$; for if $x^k = 1$ in $\mathbf{Z}_n$ then clearly x is a unit with inverse x^{k-1} and, on the other hand, every unit of $\mathbf{Z}_n$ is a $\phi(n)$-th root of 1 by Euler's Theorem. Each unit of $\mathbf{Z}_n$ is therefore a primitive kth root of 1 for some $k \geqslant 1$; this k is the multiplicative order of the unit in the group U_n.

Example 9.4. In $\mathbf{Z}_8$, the units are 1, 3, 5 and 7. Of these, 1 has order 1, and the others have order 2. Thus $\mathbf{Z}_8$ has 3 primitive square-roots of 1 and no primitive 4th root of 1. The group of units U_8 is therefore not cyclic but is the Klein 4-group (see Example 6.5 and Exercise 12 of Chapter 6).

Example 9.5. In $\mathbf{Z}_7$ the units are 1, 2, 3, 4, 5 and 6. The powers of 3 modulo 7 are $3^1 = 3$, $3^2 = 2$, $3^3 = -1$, $3^4 = -3$, $3^5 = -2$, $3^6 = 1$, so 3 is a primitive 6th root of 1 in $\mathbf{Z}_7$ and U_7 is a cyclic group of order 6.

The question whether the group U_n is cyclic or not can easily be answered for a particular value of n by direct computation of the powers of all the units. In the case when p is prime it can be proved that U_p is indeed cyclic, that is, there exists a primitive $(p-1)$-th root modulo p for each prime number p. This is quite hard. It is even harder to give a general answer to the question "when is U_n cyclic?", but some cases can be settled using the following result.

THEOREM 9C. *If m and n are positive integers and $(m, n) = 1$, then*

$$\mathbf{Z}_m \times \mathbf{Z}_n \cong \mathbf{Z}_{mn},$$

an isomorphism of rings.

Proof. The notation $\bar{r}$ is ambiguous here since we are concerned with three different rings $\mathbf{Z}_m$, $\mathbf{Z}_n$ and $\mathbf{Z}_{mn}$. We therefore write $\bar{r}_{(m)}$ and $\bar{r}_{(n)}$ for the residue classes of r in the rings $\mathbf{Z}_m$, $\mathbf{Z}_n$, respectively. The maps $r \mapsto \bar{r}_{(m)}$ and $r \mapsto \bar{r}_{(n)}$, from $\mathbf{Z}$ to $\mathbf{Z}_m$ and $\mathbf{Z}_n$, are homomorphisms of rings. It follows immediately, by the definition of the ring operations in $\mathbf{Z}_m \times \mathbf{Z}_n$ that the map $\theta : r \mapsto (\bar{r}_{(m)}, \bar{r}_{(n)})$ from $\mathbf{Z}$ to $\mathbf{Z}_m \times \mathbf{Z}_n$ is a homomorphism of rings. (For example, $r + s \mapsto ((\overline{r+s})_{(m)}, (\overline{r+s})_{(n)}) = (\bar{r}_{(m)} + \bar{s}_{(m)}, \bar{r}_{(n)} + \bar{s}_{(n)}) = (\bar{r}_{(m)}, \bar{r}_{(n)}) + (\bar{s}_{(m)}, \bar{s}_{(n)})$, so addition is preserved.) This much is true for arbitrary m and n. The kernel of θ is the set of integers r such that $\bar{r}_{(m)} = \bar{0}_{(m)}$ and $\bar{r}_{(n)} = \bar{0}_{(n)}$, that is, such that $m \mid r$ and $n \mid r$. If m and n are coprime, it follows that the kernel of θ is the set of all multiples of mn (see Theorem 5B(v)), i.e., Ker $\theta = mn\mathbf{Z}$. We now apply the first isomorphism theorem for rings (Theorem 8G) to deduce that the image of θ is isomorphic to $\mathbf{Z}/\text{Ker } \theta = \mathbf{Z}/mn\mathbf{Z} = \mathbf{Z}_{mn}$. But $\mathbf{Z}_{mn}$ has mn elements, the same number of elements as $\mathbf{Z}_m \times \mathbf{Z}_n$ in which the image is contained. Hence the image must be the whole of $\mathbf{Z}_m \times \mathbf{Z}_n$, and the theorem is proved.

COROLLARY. *If $(m, n) = 1$ then $U_m \times U_n \cong U_{mn}$, an isomorphism of groups.*

Proof. Apply Theorem 8D to obtain
$$U_{mn} = U(\mathbf{Z}_{mn}) \cong U(\mathbf{Z}_m \times \mathbf{Z}_n) \cong U(\mathbf{Z}_m) \times U(\mathbf{Z}_n)$$
$$= U_m \times U_n.$$

Example 9.6. From this corollary we can immediately deduce that if n has the prime factorization $n = p_1^{\alpha_1} p_2^{\alpha_2} \ldots p_r^{\alpha_r}$, then $U_n \cong U_{q_1} \times U_{q_2} \times \ldots \times U_{q_r}$, where q_i is the prime-power $p_i^{\alpha_i}$. For this group to be cyclic it is necessary and sufficient that each of the factors should be cyclic and that their orders should be pairwise coprime. (See Exercises 1 and 2, Chapter 6.) Thus, for example, $U_{72} \cong U_9 \times U_8$ is not cyclic since U_8 is not cyclic, as we showed in Example 9.4. Also $U_{36} \cong U_9 \times U_4$ is not cyclic because the orders of the two factors are $\phi(9) = 6$ and $\phi(4) = 2$ which are not coprime.

We observe here that Theorem 9C and its proof are intimately related to the Chinese remainder theorem (Theorem 7C). The proof is the same in both cases and it shows that when $(m, n) = 1$ the natural homomorphism $\mathbf{Z} \to \mathbf{Z}_m \times \mathbf{Z}_n$ is surjective, which is the statement that for arbitrary integers a, b there is an integer r satisfying $r \equiv a \pmod{m}$ and $r \equiv b \pmod{n}$.

We now turn to the construction of fields of fractions. We cannot apply it fruitfully to the finite rings $\mathbf{Z}_n$ because those which are integral domains are already fields. However, since $\mathbf{Z}$ is an integral domain the construction can be applied to it and the resulting field of fractions is $\mathbf{Q}$, the field of rational numbers. This field has already appeared many times in our examples, but this is the first time it has occurred in the logical exposition. Indeed, the justification for using rational numbers at all lies in Theorem 8J, which shows the existence of a field whose members are fractions m/n with $m, n \in \mathbf{Z}, n \neq 0$. (The use of the fields $\mathbf{R}$ and $\mathbf{C}$ should be similarly justified, but their construction is better suited to a course on analysis.)

Any rational number x can be written
$$x = \frac{a}{b}$$

with $b > 0$

(since $\dfrac{a}{b} = \dfrac{-a}{-b}$).

By the Euclidean property of **Z**, we have $a = bq + r$ uniquely, with $0 \leq r < b$. Hence

$$x = q + \frac{r}{b}$$

uniquely, with $q, r \in \mathbf{Z}$ and $0 \leq r < b$. We call q and r/b the integral and fractional parts of x, respectively. We shall now develop the theory of *partial fractions* whose aim is to express x as a sum of fractions of especially simple type, namely fractions with prime-power denominators.

LEMMA. *If $b = b_1 b_2$, where b_1, b_2 are coprime integers, then*

$$\frac{1}{b} = \frac{r_1}{b_1} + \frac{r_2}{b_2}$$

in **Q** *for suitably chosen integers r_1 and r_2.*

Proof. The given equation in **Q** is equivalent to the equation $1 = r_1 b_2 + r_2 b_1$ in **Z** (multiplying both sides by b). This equation holds for suitable r_1 and r_2 since b_1 and b_2 are coprime (Theorem 5B(iii)).

THEOREM 9D. *Every rational number*

$$x = \frac{a}{b}$$

can be uniquely expressed in the form

$$x = x_0 + \frac{a_1}{b_1} + \frac{a_2}{b_2} + \ldots + \frac{a_s}{b_s}, \qquad (1)$$

where $x_0, a_i, b_i \in \mathbf{Z}$, $b_1, b_2, \ldots, b_s$ are powers of distinct prime numbers, $0 < a_i < b_i$ for each i, and $b_i | b$ for each i.

Proof. The existence is an easy consequence of the lemma. We write

$$x = \frac{a}{b}$$

with $b > 0$. If $b = 1$ then x is an integer and we take $x_0 = x$. If $b > 1$ then $b = b_1 b_2 \ldots b_s$ where $b_1, \ldots, b_s$ are powers of distinct primes. Since $b_1, \ldots, b_s$ are pairwise coprime, a simple induction using the lemma gives

$$\frac{1}{b} = \frac{r_1}{b_1} + \frac{r_2}{b_2} + \ldots + \frac{r_s}{b_s}$$

for suitable integers $r_1, r_2, \ldots, r_s$. Hence

$$\frac{a}{b} = \frac{ar_1}{b_1} + \frac{ar_2}{b_2} + \ldots + \frac{ar_s}{b_s},$$

and if we now collect together the integral parts of the terms

$$\frac{ar_i}{b_i}$$

we obtain an expression of the required form with $0 \leqslant a_i < b_i$. If any of the a_i are 0, we simply omit these terms to satisfy the requirement $0 < a_i < b_i$.

The uniqueness is a little more tricky. Suppose that x has another such expression

$$x = y_0 + \frac{c_1}{d_1} + \ldots + \frac{c_t}{d_t}.$$

Then

$$\frac{a_1}{b_1} + \frac{a_2}{b_2} + \ldots + \frac{a_s}{b_s} - \frac{c_1}{d_1} - \ldots - \frac{c_t}{d_t} \in \mathbf{Z},$$

and if we add together any pair of terms whose denominators are powers of the same prime number we obtain a formula

$$\frac{u_1}{v_1} + \frac{u_2}{v_2} + \ldots + \frac{u_r}{v_r} \in \mathbf{Z}, \tag{2}$$

where $v_1, v_2, \ldots, v_r$ are powers of distinct primes, and it is easy to see that $|u_i| < v_i$ for each i. (This last fact is obvious if

$$\frac{u_i}{v_i}$$

is one of the terms

$$\frac{a_j}{b_j} \quad \text{or} \quad -\frac{c_j}{d_j}.$$

Otherwise u_i/v_i is of the form

$$\frac{g}{p^m} - \frac{h}{p^n}$$

with $0 < g < p^m$ and $0 < h < p^n$. Assuming, say, $m \leqslant n$, we have

$$\frac{u_i}{v_i} = \frac{gp^{n-m}-h}{p^n}$$

and clearly $|gp^{n-m} - h| < p^n$.) Now, in (2), let us multiply by $v_2 v_3 \ldots v_r$. Then all the fractions become integers except the first, so we have

$$\frac{u_1 v_2 v_3 \ldots v_r}{v_1} \in \mathbf{Z}, \quad \text{i.e., } v_1 \mid u_1 v_2 v_3 \ldots v_r.$$

But v_1 is coprime to $v_2 v_3 \ldots v_r$, so we deduce that $v_1 \mid u_1$. Since $|u_1| < v_1$, this implies $u_1 = 0$. Similarly, we can prove that $u_i = 0$ for all i. Thus the fractional terms of the two expressions for x must be exactly the same and hence so are the integral terms.

We can take the partial fractions process a little further by breaking up the terms of type q/p^m, where p is prime.

THEOREM 9E. *Let p be a fixed prime number. Then any positive integer q can be written uniquely in the form* $q_0 + q_1 p + q_2 p^2 + \ldots + q_n p^n$, *where* $0 \leq q_i < p$ *and* $q_n \neq 0$.

Proof. If $q < p$ we take $n = 0$, $q_0 = q$. If $q \geq p$ we have $q = q'p + q_0$, where $q' > 0$, $0 \leq q_0 < p$. Using induction on q we now see that q', being smaller than q, can be written $q' = q_1 + q_2 p + \ldots + q_n p^{n-1}$ and hence $q = q_0 + q_1 p + \ldots + q_n p^n$. The uniqueness of the expression is proved as follows. Suppose that $q_0 + q_1 p + q_2 p^2 + \ldots + q_n p^n = r_0 + r_1 p + r_2 p^2 + \ldots + r_m p^m$ with $0 \leq q_i < p$ and $0 \leq r_i < p$. Reducing modulo p we find $q_0 \equiv r_0 \pmod{p}$, whence $q_0 = r_0$ because of their restricted range. Subtracting $q_0 = r_0$ and cancelling p we obtain $q_1 + q_2 p + \ldots + q_n p^{n-1} = r_1 + r_2 p + \ldots + r_n p^{n-1}$, and a simple induction completes the proof.

THEOREM 9F. *Every rational number*

$$x = \frac{a}{b}$$

can be expressed uniquely in the form

$$x = x_0 + \frac{r_1}{p_1^{\alpha_1}} + \frac{r_2}{p_2^{\alpha_2}} + \ldots + \frac{r_k}{p_k^{\alpha_k}}, \qquad (3)$$

where x_0, r_i, p_i, α_i are integers, $p_1, p_2, \ldots, p_k$ are prime numbers (not necessarily distinct), $0 < r_i < p_i$ for all i, $\alpha_i > 0$ for all i, and the prime-powers $p_1^{\alpha_1}, p_2^{\alpha_2}, \ldots, p_k^{\alpha_k}$ are all distinct and divide the denominator b of x.

Proof. In the partial fraction decomposition for x given in Theorem 9D, each term other than x_0 is of the form a/p^n for some prime number p, with $0 < a < p^n$, and $p^n \mid b$. By Theorem 9E, a can be written $a = q_0 + q_1 p + \ldots + q_r p^r$, with $0 \leq q_i < p$ and $q_r \neq 0$. Clearly $r < n$, so on dividing by p^n we get

$$\frac{a}{p^n} = \frac{q_0}{p^n} + \frac{q_1}{p^{n-1}} + \ldots + \frac{q_r}{p^{n-r}}, \tag{4}$$

which is of the required form. If we do this for each term and add the results we can express x in the required form (3). The expression is unique, for if we collect together all terms whose denominators are powers of the same prime and add them, we obtain a decomposition

$$x = x_0 + \frac{a_1}{b_1} + \ldots + \frac{a_s}{b_s}$$

of the type given in Theorem 9E. This decomposition is unique, so the a_i/b_i are uniquely determined. But each of these has a unique decomposition of the form (4), so the terms of (3) are uniquely determined.

Example 9.7. The rational number

$$\frac{14}{135} = \frac{14}{3^3 \cdot 5}$$

can be written, according to Theorem 9F in the form

$$a + \frac{b}{5} + \frac{c}{3} + \frac{d}{3^2} + \frac{e}{3^3}$$

where a, b, c, d, e are integers and $0 \leq b < 5, 0 \leq c < 3, 0 \leq d < 3$ and $0 \leq e < 3$. To find the numerators, multiply by $3^3 \cdot 5$ to obtain $14 = 3^3 \cdot 5a + 3^3 b + 3^2 \cdot 5c + 3 \cdot 5d + 5e$. Modulo 5, this gives $27b \equiv 14 \pmod 5$, i.e., $2b \equiv 4 \pmod 5$, whence $b = 2$ (since $0 \leq b < 5$). Substituting back, we find $-40 = 3^3 \cdot 5a + 3^2 \cdot 5c + 3 \cdot 5d + 5e$, whence $-8 = 27a + 9c + 3d + e$ with c, d, e in the range $\{0, 1, 2\}$. Modulo 3 this gives $e \equiv -8 \equiv 1$, so $e = 1$. Hence $-3 = 9a + 3c + d$, which gives $d = 0$. Continuing in this way, we find $c = 2, a = -1$ and so we may deduce

$$\frac{14}{135} = -1 + \frac{2}{5} + \frac{2}{3} + \frac{1}{27}. \tag{5}$$

Note the logic of this assertion: because of the theorem we know there is a decomposition of the required form; assuming 14/135 is expressed in this form we have proved that the only possible values of the numerators are the ones given; therefore (5) is true. The only reason one would have for checking it is to make sure that no arithmetical errors have been made on the way. However, it is a common mistake to try and apply the same argument to decompose a number into a form which is not actually possible. For example, if one misunderstood the theorems, one might think that, because 7/45 has integral part 0, it can be written in the form

$$\frac{7}{45} = \frac{a}{3} + \frac{b}{9} + \frac{c}{5}$$

with $0 \leqslant a < 3, 0 \leqslant b < 3$ and $0 \leqslant c < 5$. If we assume this and multiply by 45, we obtain $7 = 15a + 5b + 9c$. Modulo 3 this gives $5b \equiv 7 \pmod{3}$, which implies $b = 2$ and $-3 = 15a + 9c$. Modulo 5 this gives $9c \equiv -3 \pmod 5$, and modulo 9 it gives $15a \equiv -3 \pmod 9$. These imply $c = 3$ and $a = 1$. Thus one might deduce that

$$\frac{7}{45} = \frac{1}{3} + \frac{2}{9} + \frac{3}{5},$$

which is false. The point is that 7/15 *cannot* be written in the proposed form, so any information one can deduce about a, b and c is valueless.

Exercises.

1. Prove that the group of units of $\mathbf{Z}_n$ is cyclic for $n = 2, 3, 4, 5, 6$ and 7. Show that it is not cyclic for $n = 15$.

2. Find a generator for the multiplicative group of the field $\mathbf{Z}_{23}$.

3. Find the least positive integer n such that $27^n \equiv 1 \pmod{77}$.

4. Prove that $n^{22} \equiv n^2 \pmod{25}$ for all $n \in \mathbf{Z}$.

5. Prove that $n^{61} \equiv n \pmod{1{,}001}$ for all $n \in \mathbf{Z}$.

6. Prove that $n^8 \equiv n^2 \pmod{252}$ for all $n \in \mathbf{Z}$.

7. Prove that $n^{17} \equiv n \pmod{8160}$ for all odd integers n. For which even integers n is the congruence true? Find the largest integer N such that, for all $n \in \mathbf{Z}$, $n^{17} \equiv n \pmod{N}$.

8. Show that the integers 2, 4, 6 and 8 form a group under multiplication modulo 10. What is its identity element? Is it cyclic?

9. Let p be an odd prime number and suppose that $a^p + b^p \equiv 0 \pmod p$. Deduce that $a^p + b^p \equiv 0 \pmod{p^2}$.

10. Express 379/1200 as a sum of fractions with prime-power denominator.

11. Find the full decomposition of 29/180 into partial fractions, as in Theorem 9F.

12. Let R be a commutative ring and let $x \in R$. Prove that x is a unit of R if and only if the principal ideal xR generated by x is the whole of R.

13. Prove that each ideal of $\mathbf{Z}_n$ is a principal ideal of the form $d\mathbf{Z}_n$, where $d \mid n$. Show that the corresponding quotient ring is isomorphic to $\mathbf{Z}_d$.

14. Prove that if R is a subring of $\mathbf{Q}$ then the field of fractions of R is isomorphic to $\mathbf{Q}$. (Hint: use the universal property of the field of fractions and the fact that $R \supset \mathbf{Z}$.)

15. Let p be a fixed prime number and let L_p be the subset of $\mathbf{Q}$ consisting of all fractions a/p^r for $a, r \in \mathbf{Z}$, $r \geq 0$. Prove that L_p is a subring of $\mathbf{Q}$ and that each non-zero ideal of L_p is of the form dL_p where d is a positive integer not divisible by p.

16. Let P be a set of prime numbers, and let L_P denote the set of all rational numbers whose denominators are products of powers of prime numbers in P. Show that L_P is a subring of $\mathbf{Q}$. Show, conversely, that every subring of $\mathbf{Q}$ is equal to L_P for some set P of primes. (Hint: for the converse, show that if a subring R of $\mathbf{Q}$ contains a rational number whose denominator is divisible by a given prime p but whose numerator is not divisible by p, then $1/p \in R$ and hence $R \supset L_p$.)

17. Prove that every rational number r, with $0 < r < 1$ can be written in the form
$$r = \frac{1}{n_1} + \frac{1}{n_2} + \ldots + \frac{1}{n_k},$$
where $n_1, n_2, \ldots, n_k$ are integers and $n_1 < n_2 < \ldots < n_k$. Can this be done for every positive rational number? (Do not be misled! This has nothing to do with partial fractions, but is thrown in for good measure.)

CHAPTER 10

Rings of Polynomials

A polynomial is, essentially, an expression of the form
$$a_0 + a_1 X + a_2 X^2 + \ldots + a_n X^n, \tag{1}$$
but some care is needed to make the definition precise. We must make clear what coefficients a_i are allowed (for example, can we use X as a coefficient?) and we must be able to tell from the definition whether $0X^2$ is the same polynomial as $0X^4$ and whether $2X + X^2$ is the same polynomial as $X^2 + 2X$.

We start by specifying a commutative ring R from which the coefficients are to be chosen and a symbol X, not a member of R. We define a *polynomial in X with coefficients in R* to be a formal expression

$$\sum_{i=0}^{\infty} a_i X^i \quad \text{(or, for brevity, } \Sigma a_i X^i\text{),}$$

where the coefficients $a_0, a_1, a_2, \ldots$ are members of R and all but a finite number of them are 0. In this definition the Σ does not (yet) signify addition. Nor does X stand for a variable element of R; its purpose is simply to label the various terms in a convenient way for later computations. Two polynomials $\Sigma a_i X^i$ and $\Sigma b_i X^i$ are equal if and only if $a_i = b_i$ for all $i \in \mathbf{N}$. The set of all such polynomials is denoted by $R[X]$, and our next task is to endow this set with a ring structure.

The *sum* of two polynomials $\Sigma a_i X^i$ and $\Sigma b_i X^i$ is defined to be $\Sigma c_i X^i$, where $c_i = a_i + b_i$ for all $i \in \mathbf{N}$; this sum is again a polynomial because $c_i = a_i + b_i \in R$ for each i and only a finite number of the c_i are different from 0. Thus we have a binary operation + defined on $R[X]$ which clearly makes $R[X]$ an Abelian group. Its zero element is the polynomial all of whose coefficients are 0 (we denote this polynomial also by 0). The negative of the polynomial $\Sigma a_i X^i$ is the polynomial $\Sigma(-a_i)X^i$, which we now also write as $-\Sigma a_i X^i$. The associative and commutative laws for addition of polynomials follow immediately from the same laws in R.

The *product* of two polynomials $\Sigma a_i X^i$ and $\Sigma b_i X^i$ is defined to be the polynomial $\Sigma d_i X^i$, where
$d_n = a_0 b_n + a_1 b_{n-1} + \ldots + a_i b_{n-i} + \ldots + a_n b_0$. We observe first that

123

$d_n \in R$ for all n, and also that all but a finite number of the d_n are 0 (for $\exists N$ such that $a_i = b_i = 0$ for $i > N$, and it is clear that $d_n = 0$ for $n > 2N$). Thus we have a well-defined binary operation of multiplication on $R[X]$ and we claim that, with the addition defined above, $R[X]$ is now a commutative ring. There are a number of laws still to be checked, and the obvious ones are (i) the commutative law of multiplication, which follows from the equation

$$d_n = a_0 b_n + a_1 b_{n-1} + \ldots + a_n b_0 = b_0 a_n + b_1 a_{n-1} + \ldots + b_n a_0,$$

and (ii) the fact that $R[X]$ has an identity element 1 which is the polynomial $\Sigma a_i X^i$ in which $a_0 = 1$ and all other a_i are 0. There remain the associative law of multiplication and the distributive law, neither of which is immediately obvious. For the associative law, write $a(X) = \Sigma a_i X^i$, $b(X) = \Sigma b_i X^i$, $c(X) = \Sigma c_i X^i$, and let $d(X) = a(X)b(X)$, $e(X) = b(X)c(X)$. We have to show that $d(X)c(X) = a(X)e(X)$. Now $d(X) = \Sigma d_i X^i$, where

$$d_n = a_0 b_n + a_1 b_{n-1} + \ldots + a_n b_0 = \sum_{i+j=n} a_i b_j.$$

Hence, by definition, $d(X)c(X) = \Sigma g_i X^i$ where

$$g_m = d_0 c_m + d_1 c_{m-1} + \ldots + d_m c_0$$

$$= \sum_{n+k=m} d_n c_k$$

$$= \sum_{n+k=m} \left(\sum_{i+j=n} a_i b_j \right) c_k$$

$$= \sum_{i+j+k=m} a_i b_j c_k.$$

The symmetry of this formula implies that if we compute the coefficients of $a(X)e(X)$ we shall get the same result and so the associative law is true in $R[X]$. Its truth depends heavily on the associative and distributive laws in R. The distributive law is proved by a similar computation of the coefficients of $a(X) \{b(X) + c(X)\}$ and $a(X)b(X) + a(X)c(X)$ using the definitions of sum and product in $R[X]$; we leave this as an exercise, and urge the reader to write it out completely because the ring laws will be used for polynomials without further question, and he should be convinced of the truth of all of them.

The properties of the commutative ring $R[X]$ will be investigated systematically, but first we introduce some simplifying notations. A polynomial $\Sigma a_i X^i$ in which $a_i = 0$ for all $i \geqslant 1$ will be denoted simply by a_0. This may seem rash, because a_0 also denotes an element of R. However the sum and product of a_0 and b_0, say, as polynomials, is the same as their sum and product in R (check this from the definitions!).

Thus we may identify R with the set of all polynomials of this simple type and so treat R as a subring of $R[X]$.

Any non-zero polynomial $\Sigma a_i X^i$ has at least one non-zero coefficient and therefore a *last* non-zero coefficient, that is, a coefficient $a_n \neq 0$ such that $a_i = 0$ for all $i > n$. This coefficient is called the *leading coefficient* (although it is really the trailing coefficient in our notation), the term $a_n X^n$ is called the *leading term* and the integer n is called the *degree* of the polynomial. The polynomial 0 does not have a degree in this sense and one must guard against supposing that its degree is 0. The polynomials of degree 0 are those with $a_0 \neq 0$ and $a_i = 0$ for $i > 0$. Thus the subring R of $R[X]$ consists of the polynomials of degree 0 together with the zero polynomial. We write $\deg(a(X))$ for the degree of the polynomial $a(X)$ and, for reasons which will be apparent later, we conventionally assign the degree $-\infty$ to the zero polynomial.

The polynomial of degree 1 whose coefficients are $a_0 = 0, a_1 = 1$, $a_i = 0$ for $i > 1$, will be denoted simply by X. Applying the definition of multiplication to this polynomial we find that $X^m = XX \ldots X$ (m factors) is the polynomial with coefficients $a_m = 1$, $a_i = 0$ for $i \neq m$. Similarly for $a \neq 0$ in R the polynomial aX^m has coefficients $a_m = a$, and $a_i = 0$ for $i \neq m$. Thus, if a polynomial $\Sigma a_i X^i$ has degree n it is actually the sum in $R[X]$ of the polynomials $a_0, a_1 X, a_2 X^2, \ldots, a_n X^n$ and so can be written in the form (1). The summation sign in the definition of polynomials now appears in its true light, and so does the notation $a_i X^i$ for its summands. Of course, since addition in $R[X]$ is commutative the polynomial $\Sigma a_i X^i$ of degree n can also be written $a_n X^n + a_{n-1} X^{n-1} + \ldots + a_1 X + a_0$ or in many other ways.

THEOREM 10A. *Let $a(X), b(X)$ be non-zero polynomials in $R[X]$.*

Then
(i) $\deg(a(X) + b(X)) \leqslant \max\{\deg(a(X)), \deg(b(X))\}$;
(ii) $\deg(a(X)b(X)) \leqslant \deg(a(X)) + \deg(b(X))$.

If R is an integral domain then

$$\deg(a(X)b(X)) = \deg(a(X)) + \deg(b(X))$$

and hence $R[X]$ is also an integral domain.

Proof. Let $m = \deg(a(X))$, $n = \deg(b(X))$ where, as usual, $a(X) = \Sigma a_i X^i$ and $b(X) = \Sigma b_i X^i$. Then $a_i = 0$ for $i > m$, and $b_i = 0$ for $i > n$. Hence $a_i + b_i = 0$ for $i > \max\{m, n\}$, and (i) follows. Similarly if $a(X)b(X) = \Sigma d_i X^i$ then

$$d_r = \sum_{i+j=r} a_i b_j = 0$$

if $r > m + n$, and (ii) follows. However, if R is an integral domain, then

the leading coefficient of the product is
$d_{m+n} = a_0 b_{m+n} + \ldots + a_m b_n + \ldots + a_{m+n} b_0 = a_m b_n$ and this is not zero since $a_m \neq 0, b_n \neq 0$. Hence $a(X)b(X)$ is not the zero polynomial and $R[X]$ is an integral domain. (The formulae remain true when $a(X)$ or $b(X)$ is 0 if such expressions as $-\infty + n$ are interpreted in an obvious way.)

Because of this theorem we shall, from now on, confine our attention almost entirely to polynomials with coefficients in an integral domain D. For some results we shall even want to assume that the coefficient domain is a field and we shall then denote it by F. Note however that if F is a field, we cannot deduce that $F[X]$ is a field. In fact it never is, because of the following result.

THEOREM 10B. *If D is an integral domain then the units of the polynomial ring $D[X]$ are the units of D. In particular, if F is a field, the units of $F[X]$ are precisely the polynomials of degree 0.*

Proof. It is clear that the units of D remain units when considered as members of $D[X]$. Conversely, suppose that $a(X)$ is a unit of $D[X]$. Then $a(X)b(X) = 1$ for some polynomial $b(X) \in D[X]$. Hence $\deg(a(X)) + \deg(b(X)) = 0$ by Theorem 10A, since D is an integral domain. It follows that both $a(X)$ and $b(X)$ have degree 0 and are therefore elements of the subring D. Since their product is 1, they are both units of D. When $D = F$ is a field, these units are all the non-zero elements of F, i.e., all the polynomials of degree 0.

For polynomial rings $F[X]$ over a *field* F there is a very strong analogy with the ring **Z**, based on the following "Euclidean property", which should be compared with Theorem 3C.

THEOREM 10C. *Let F be a field and let $a(X), b(X)$ be polynomials in $F[X]$ with $b(X) \neq 0$. Then there exist polynomials $q(X), r(X)$ in $F[X]$ such that $a(X) = b(X)q(X) + r(X)$ and $\deg(r(X)) < \deg(b(X))$. Furthermore, $q(X)$ and $r(X)$ are unique subject to these properties.*

Proof. Amongst all polynomials of the form $a(X) - b(X)q(X)$, for arbitrary $q(X)$, let $r(X)$ be one of least possible degree. Recalling our convention that the zero polynomial has degree $-\infty$, which is construed as less than all natural numbers, $r(X)$ will be the zero polynomial if 0 is amongst the polynomials of the given form; otherwise the set of all the degrees of these polynomials is a non-empty set of natural numbers and so has a least member, which guarantees the existence of $r(X)$. We now show that $\deg(r(X)) < \deg(b(X))$. For suppose not, and let the leading terms of $r(X)$ and $b(X)$ be respectively $r_n X^n$ and $b_m X^m$, with $n \geq m$. Since $b_m \neq 0$ and F is a field, the polynomial $s(X) = b_m^{-1} r_n X^{n-m}$ is in $F[X]$, and $b(X)s(X)$ has leading term $r_n X^n$, the same as that of $r(X)$. Hence the polynomial $t(X) = r(X) - b(X)s(X)$ has smaller degree than $r(X)$. But $r(X) = a(X) - b(X)q(X)$ for some $q(X)$, so $t(X) = a(X) - b(X)\{q(X) + s(X)\}$ is again of the same form and has smaller degree than $r(X)$, a contradiction. This proves that deg $(r(X)) <$ deg $(b(X))$. The uniqueness of $q(X)$ and $r(X)$ now follows easily: if $a(X) = b(X)q_1(X) + r_1(X)$ with deg $(r_1(X)) <$ deg $(b(X))$, then

$$b(X) \{q(X) - q_1(X)\} = r_1(X) - r(X).$$

By Theorem 10A, the right-hand side has degree less than deg $(b(X))$, but the left-hand side has degree at least deg $(b(X))$ unless $\{q(X) - q_1(X)\} = 0$. We deduce that $q(X) = q_1(X)$ and hence that $r(X) = r_1(X)$.

Next, we prove the analogue of Theorem 4C on subgroups of **Z**, which was the basis of factorization theory in **Z**. We call a polynomial *monic* if its leading coefficient is 1, and observe that any non-zero polynomial over a field can be made monic by multiplying by a unit, namely, the inverse of its leading coefficient.

THEOREM 10D. *(Principal Ideal Theorem.) Let F be a field. Then every ideal of $F[X]$ is a principal ideal $d(X) F[X]$ for some $d(X) \in F[X]$. The generator $d(X)$ can be chosen to be either 0 or a monic polynomial and it is then unique.*

Proof. Let I be any ideal of $F[X]$. If $I = \{0\}$ then we may take $d(X) = 0$ and this is clearly unique. Otherwise, I contains some non-zero polynomials and we choose one of least possible degree. If we multiply this polynomial by a unit it remains in I and its degree is unchanged, so we may choose it to be monic. Call this polynomial $d(X)$. Clearly $d(X)F[X] \subset I$ and it remains to show that every $a(X) \in I$ is of the form $a(X) = d(X)q(X)$ for some $q(X)$. By Theorem 10C, we know that $a(X) = d(X)q(X) + r(X)$, where deg $(r(X)) <$ deg $(d(X))$. But $r(X) = a(X) - d(X)q(X) \in I$, so we arrive at a contradiction unless $r(X) = 0$. (Recall that $d(X)$ was a *non-zero* element of least degree in I.) Thus we must have $r(X) = 0$ and $a(X) = d(X)q(X)$, as required. The uniqueness of $d(X)$ follows because, if $d_1(X)$ is another monic generator of I, then $d_1(X) = d(X)q(X)$ and $d(X) = d_1(X)q_1(X)$, so $d(X) = d(X)q(X)q_1(X)$. By the degree formula, we see that $q(X)q_1(X)$ has degree 0 and hence $q(X)$ and $q_1(X)$ are both units. But $d_1(X)$ and $d(X)$ are both monic, so it follows that $q(X) = q_1(X) = 1$, whence $d_1(X) = d(X)$.

The way is now clear for proving analogues of all the theorems of Chapter 5. Indeed, we could prove simultaneously the corresponding theorems for $\mathbf{Z}$ and $F[X]$ by introducing the idea of "Euclidean domain", which is an integral domain with a suitable Euclidean property. However, since we shall only be concerned with these two examples of such structures, the axiomatic device is not worthwhile. Instead, we shall state the definitions and theorems for polynomial rings and give sufficient indication of the proofs to enable the reader to complete them. To save space we shall usually omit the symbol X in polynomials and write $a, b, \ldots \in F[X]$ to mean $a(X), b(X), \ldots \in F[X]$.

Divisibility in a polynomial ring $D[X]$ is defined as usual: $a \mid b$ means that $\exists c \in D[X]$ such that $b = ac$. The reader will easily verify the following elementary properties, the last two of which depend on D being an integral domain.

(i) $a \mid 0$ for all $a \in D[X]$;

(ii) $0 \mid a \Leftrightarrow a = 0$;

(iii) $a \mid b$ and $b \mid c \Rightarrow a \mid c$;

(iv) $a \mid b$ and $a \mid c \Rightarrow a \mid bp + cq$ for all $p, q \in D[X]$;

(v) $a \mid 1$ in $D[X] \Leftrightarrow a$ is a unit of D;

(vi) $a \mid b$ and $b \mid a \Leftrightarrow a = bu$, where u is a unit of D.

For polynomials $a, b \in D[X]$, where D is an integral domain and a, b are not both zero, we say that $d \in D[X]$ is a *greatest common divisor* of a and b if

(i) $d \mid a$ and $d \mid b$;

(ii) $c \mid a$ and $c \mid b \Rightarrow c \mid d$;

(iii) d is monic.

As in the case of integers, the third condition is added for convenience to ensure that d, if it exists, is unique. For if d and d_1 both satisfy the three conditions then $d \mid d_1$ and $d_1 \mid d$, whence $d = d_1 u$, for some unit u of D. Since d and d_1 are monic, we must have $d = d_1$. We cannot, however, prove *existence* of greatest common divisors in this generality and, for the present, we only deal with the case when D is a field.

THEOREM 10E.   *Let F be a field and let $a, b \in F[X]$ be not both zero. Then a and b have a unique greatest common divisor d in $F[X]$ and d can be written in the form $d = ap + bq$ for suitable polynomials $p, q \in F[X]$.*

Proof.   Let I be the set of all polynomials of the form $ap + bq$, where a, b are the given, fixed, polynomials and p, q are arbitrary polynomials. Then I is an *ideal* of $F[X]$. (We leave the reader to check this.) By Theorem 10D, this ideal is principal, generated by a monic polynomial d. (Note that $I \neq \{0\}$ since $a \in I$ and $b \in I$.) Since $d \in I$, we have $d = ap + bq$ for suitable $p, q \in F[X]$. Hence, if $c \mid a$ and $c \mid b$, we have $c \mid d$. Finally, since $I = dF[X]$ and $a, b \in I$, we have $d \mid a$ and $d \mid b$. The uniqueness of d has already been proved.

As in **Z**, we write $d = (a, b)$ for the greatest common divisor, and say that a and b are *coprime* if $(a, b) = 1$. We also adopt the convention that $(0,0) = 0$ (although this is not a monic polynomial).

THEOREM 10F. *Let F be a field. Then*

(i) $(ac, bc) = (a, b)c$ *for all* $a, b, c \in F[X]$ *with c monic*;

(ii) *if* $d = (a, b) \neq 0$ *in* $F[X]$, *then* $a = da_1$, $b = db_1$ *where* a_1 *and* b_1 *are coprime*;

(iii) a *and* b *are coprime in* $F[X]$ $\Leftrightarrow$ $(\exists p, q \in F[X])(ap + bq = 1)$;

(iv) *if* $a \mid bc$ *in* $F[X]$ *and* $(a, b) = 1$, *then* $a \mid c$;

(v) *if* $a \mid c$ *and* $b \mid c$ *in* $F[X]$ *and* $(a, b) = 1$, *then* $ab \mid c$;

(vi) *if* $a = bq + r$ *in* $F[X]$ *then* $(a, b) = (b, r)$.

Proof. The reader should check that the proof of Theorem 5B is valid in the context of polynomials with an occasional minor modification when a condition such as $c > 0$ is replaced by the condition "c is monic".

The last statement (vi) of this theorem is, in the case of **Z**, the basis for Euclid's algorithm, and we can prove by the same argument that Euclid's algorithm can be used to compute greatest common divisors of polynomials (over a field). Of course, division of polynomials is harder to perform than division of integers, so the algorithm is more troublesome, but the principle is exactly the same. Given polynomials $a_0, a_1 \in F[X]$ with $a_1 \neq 0$ we write

$$a_0 = a_1 q_1 + a_2$$
$$a_1 = a_2 q_2 + a_3$$
$$\cdot$$
$$\cdot$$
$$\cdot$$
$$a_{n-2} = a_{n-1} q_{n-1} + a_n$$
$$a_{n-1} = a_n q_n$$

where, at each step, $\deg(a_i) < \deg(a_{i-1})$. Eventually we must have

$a_{n+1} = 0$ since the degrees cannot decrease indefinitely. If $a_n \neq 0$, then a_n is the greatest common divisor, after being made monic by multiplying by a unit, because

$$(a_0, a_1) = (a_1, a_2) = \ldots = (a_{n-1}, a_n) = (a_n, 0) = a_n u,$$

where u is the inverse of the leading coefficient of a_n.

Example 10.1. Let $F = \mathbf{Q}$ and let

$$a_0(X) = 12X^4 - 9X^3 - 14X^2 + 15X - 10$$

and

$$a_1(X) = 4X^4 + 9X^3 - 7X^2 + 6X.$$

To compute the g.c.d. of a_0 and a_1 we do the necessary long divisions in easy stages, the remainder being not always of least possible degree. We also multiply through by appropriate units (in fact non-zero integers) to avoid some fractions. Neither of these modifications of the algorithm affects the result.

$$a_0 = 3a_1 + a_2, \quad \text{where } a_2 = -36X^3 + 7X^2 - 3X - 10;$$

$$9a_1 = (-X)a_2 + a_3, \quad \text{where } a_3 = 88X^3 - 66X^2 + 44X$$
$$= 22(4X^3 - 3X^2 + 2X);$$

$$a_2 = -\frac{9}{22}a_3 + a_4, \text{ where } a_4 = -20X^2 + 15X - 10$$
$$= 5(-4X^2 + 3X - 2);$$

$$a_3 = \left(-\frac{22}{5}X\right)a_4.$$

Hence

$$(a_0, a_1) = (a_1, a_2) = (9a_1, a_2) = (a_2, a_3) = (a_3, a_4) = (a_4, 0)$$

$$= X^2 - \frac{3}{4}X + \frac{1}{2}.$$

To express this in the form $a_0 p + a_1 q$ we reverse the computation:

$$a_4 = a_2 + \frac{9}{22}a_3 = a_2 + \frac{9}{22}(9a_1 + Xa_2) = \frac{81}{22}a_1 + \left(1 + \frac{9}{22}X\right)a_2$$

$$= \frac{81}{22}a_1 + \left(1 + \frac{9}{22}X\right)(a_0 - 3a_1)$$

$$= \left(1 + \frac{9}{22}X\right)a_0 + \left(\frac{15}{22} - \frac{27}{22}X\right)a_1.$$

The monic g.c.d. is then obtained by dividing by -20.

Example 10.2. Let $F = \mathbf{Z}_5$, and let $a_0 = 2X^3 + 4X + 1$ and $a_1 = 3X^3 + 2X^2 + 3X + 2$. Here we have omitted bars on the coefficients, which are not integers but residue classes modulo 5. However, we can compute with integers provided we remember that $2 = -3 = 7$ etc. Since the inverses of 1, 2, 3, 4 are 1, 3, 2, 4, the working is much easier. Of course, the algorithm is justified by the fact that $\mathbf{Z}_5$ is a *field*.

$a_0 = 4a_1 + a_2$, where $a_2 = 2X^2 + 2X + 3$;
$a_1 = 4Xa_2 + a_3$, where $a_3 = 4X^2 + X + 2$;
$a_2 = 3a_3 + a_4$, where $a_4 = 4X + 2$;
$a_3 = Xa_4 + a_5$, where $a_5 = 4X + 2$;
$a_4 = a_5$.

Hence the g.c.d. is $d = 4a_5 = X + 3$. Working back we get

$d = 4a_4 = 4a_2 + 3a_3 = 4a_2 + 3(a_1 + Xa_2) = (4 + 3X)a_2 + 3a_1$
$= (4 + 3X)(a_0 + a_1) + 3a_1 = (4 + 3X)a_0 + (7 + 3X)a_1$.

The analogue of "prime number" for the ring $F[X]$ is "monic irreducible polynomial", defined as follows. The polynomial $p \in F[X]$ is *irreducible* if

(i) $p \neq 0$,
(ii) p is not a unit, and
(iii)   every divisor of p in $F[X]$ is of the form u or up, where u is a unit.

Since the units of $F[X]$ are precisely the polynomials of degree zero, it is easy to see that a polynomial is irreducible if and only if it has degree at least 1 and cannot be written as the product of two polynomials of degree $\geqslant 1$. Two irreducible polynomials are *associated* if one is a unit multiple of the other. In each class of associated irreducible polynomials there is a unique *monic* irreducible polynomial (which will play the same role as the positive prime number n chosen from the pair $\pm n$). Using properties of greatest common divisors as in Theorem 5C the reader should have no difficulty in proving:

THEOREM 10G. *Let F be a field. Then*

(i)   *if p, q are monic irreducible polynomials in $F[X]$ and if $p \mid q$, then $p = q$;*
(ii)   *if $a, p \in F[X]$, if p is irreducible and if $p \nmid a$, then $(p, a) = 1$;*

(iii) *if a polynomial $a \in F[X]$ is not zero, is not a unit and is not irreducible, then $\exists b, c \in F[X]$ such that $a = bc$, $0 < \deg(b) < \deg(a)$ and $0 < \deg(c) < \deg(a)$;*

(iv) *if p is an irreducible polynomial and $p \mid a_1 a_2 \ldots a_n$ in $F[X]$, then $p \mid a_i$ for at least one value of i.*

Armed with this information we can now prove a unique factorization theorem for polynomials.

THEOREM 10H. *Let F be a field. Then*

(i) *every non-zero polynomial $a \in F[X]$ can be factorized in the form $a = up_1 p_2 \ldots p_r$, where u is a unit, each p_i is a monic irreducible polynomial, and $r \geq 0$;*

(ii) *if $a = vq_1 q_2 \ldots q_s$ is another factorization with v a unit and each q_j a monic irreducible polynomial, then $u = v$, $r = s$ and $q_1, q_2, \ldots, q_s$ are a permutation of $p_1, p_2, \ldots, p_r$.*

Proof. (i) Since every non-zero polynomial is a unit multiple of a monic polynomial, we can assume that a is monic. If a is a unit, we may take $u = 1, r = 0$. If a is irreducible, we may take $u = 1, r = 1, p_1 = a$. Otherwise, by Theorem 10G, $a = bc$ where $0 < \deg(b) < \deg(a)$ and $0 < \deg(c) < \deg(a)$. Since a is monic we can choose b and c both monic. Induction on $\deg(a)$ now gives the existence of the required factorization as in the case of **Z**. (See Theorem 5D.)

(ii) Suppose that $a = up_1 p_2 \ldots p_r = vq_1 q_2 \ldots q_s$. Clearly $u = v$ is the leading coefficient of a, so we may assume that a is monic and $u = v = 1$. If a has degree 0 then $a = 1$ and clearly $r = s = 0$. Thus we may assume that $a = p_1 p_2 \ldots p_r = q_1 q_2 \ldots q_s$ with $r \geq 1, s \geq 1$. We cannot arrange the irreducibles in ascending order, as we did for integers, but the argument is essentially the same. We use induction on the degree of a, the case $\deg(a) = 0$ having been dealt with already. We have $p_1 \mid q_1 q_2 \ldots q_s$, so $p_1 \mid q_j$ for some j, and therefore $p_1 = q_j$, by Theorem 10G. Since $F[X]$ is an integral domain (Theorem 10A) we may cancel p_1 and q_j to obtain $b = p_2 p_3 \ldots p_r = q_1 \ldots q_{j-1} q_{j+1} \ldots q_s$. But $\deg(b) < \deg(a)$, so by induction hypothesis, $q_1, \ldots, q_{j-1}, q_{j+1}, \ldots, q_s$ are a permutation of $p_2, p_3, \ldots, p_r$, and since $p_1 = q_j$, it follows that $q_1, q_2, \ldots, q_s$ are a permutation of $p_1, p_2, \ldots, p_r$.

Example 10.3. The factorization of a polynomial into irreducible factors depends very much on the field from which the coefficients are to be chosen. For example, the polynomial $X^4 - 4$ can be considered as a member of $\mathbf{C}[X], \mathbf{R}[X]$ or $\mathbf{Q}[X]$. In $\mathbf{C}[X]$ its factorization is $(X - \sqrt{2})(X + \sqrt{2})(X - i\sqrt{2})(X + i\sqrt{2})$; these factors are necessarily irreducible because they are of degree 1. In $\mathbf{R}[X]$ we have

$X^4 - 4 = (X - \sqrt{2})(X + \sqrt{2})(X^2 + 2)$ and all these factors are irreducible in $\mathbf{R}[X]$. In $\mathbf{Q}[X]$ we cannot do better than $X^4 - 4 = (X^2 - 2)(X^2 + 2)$. These assertions will be easier to justify later. However, we can be sure already that in each one of these polynomial rings, the factorisation into irreducible factors is unique.

One aspect of polynomials which is entirely lacking in the ring of integers is that polynomials can be used to define functions. If R is any commutative ring and if $a(X) \in R[X]$, then $a(X) = a_0 + a_1 X + \ldots a_n X^n$ for some $a_i \in R$. If we substitute for X in this formula a member x of R, we obtain an expression $a_0 + a_1 x + \ldots + a_n x^n$ which can be calculated in R using its ring operations, and the result is a member of R, denoted by $a(x)$. Thus the polynomial $a(X)$ determines a function $\alpha : x \mapsto a(x)$ from R to R, and we call this a *polynomial function*. The notation tempts us to denote this function by a since its value at x is $a(x)$. However, this must be avoided because *different polynomials may determine the same polynomial function.*

Example 10.4. Let $R = \mathbf{Z}_p$, where p is a prime number. In the polynomial ring $\mathbf{Z}_p[X]$, X^p and X are different polynomials since they even have different degrees. But the functions from $\mathbf{Z}_p$ to $\mathbf{Z}_p$ determined by them are $x \mapsto x^p$ and $x \mapsto x$, and these are the same function since $x^p = x$ for all $x \in \mathbf{Z}_p$, by Theorem 9B, Corollary 2.

Put another way, this warning says that a polynomial $a(X) \in R[X]$ may satisfy the "identity" $a(x) = 0$ for all $x \in R$ without being the zero polynomial (for example, the polynomial $a(X) = X^p - X$ when $R = \mathbf{Z}_p$). Thus the often-used argument "$a(x) = b(x)$ for all x, therefore we may equate the coefficients of $a(X)$ and $b(X)$" is not valid in general. It is not even valid when R is a field, as Example 10.4 shows. However we shall soon show that it *is* valid when R is an infinite field.

THEOREM 10J. *(Universal property of polynomial rings.) Let R, S be commutative rings, let $f : R \to S$ be a homomorphism of rings, and let $x \in S$. Then there is a unique homomorphism of rings $R[X] \to S$ such that (i) $r \mapsto f(r)$ for all $r \in R$ and (ii) $X \mapsto x$. This homomorphism is given by $a(X) \mapsto a^*(x)$ for all polynomials $a(X) \in R[X]$, where the polynomial $a^*(X) \in S[X]$ is obtained from $a(X)$ by applying f to all its coefficients.*

Proof. Suppose that a homomorphism F exists from $R[X]$ to S satisfying (i) and (ii). Writing r^* for $f(r)$ we see that F must send rX^n to $r^* x^n$, since it preserves multiplication. Hence F must send $\Sigma a_i X^i$ to $\Sigma a_i^* x^i$, since it also preserves addition. This shows that F is unique and is given by $a(X) \mapsto a^*(x)$. To prove existence, we let F be the map from $R[X]$ to S given by $a(X) \mapsto a^*(x)$. This is well-defined and satisfies conditions (i) and (ii). It preserves addition and multiplication *because the operations on $R[X]$ were defined precisely with this in mind.* To see this, let $a(X) = \Sigma a_i X^i$, $b(X) = \Sigma b_j X^j$. Then $a(X) + b(X)$ is, by definition, the polynomial $\Sigma(a_i + b_i)X^i$ which is mapped by F to $\Sigma(a_i + b_i)^* x^i = \Sigma(a_i^* + b_i^*)x^i$. The ring laws in S imply that this is equal to $\Sigma a_i^* x^i + \Sigma b_i^* x^i = a^*(x) + b^*(x)$. Thus F preserves addition. Similarly, $a(X)b(X) = c(X)$, where $c(X) = \Sigma c_n X^n$ and $c_n = a_0 b_n + a_1 b_{n-1} + \ldots + a_n b_0$. This polynomial is mapped by F to $c^*(x) = \Sigma c_n^* x^n \in S$. But $c_n^* = a_0^* b_n^* + a_1^* b_{n-1}^* + \ldots + a_n^* b_0^*$, since f is a ring homomorphism. It follows that

$$a^*(x)b^*(x) = (\Sigma a_i^* x^i)(\Sigma b_j^* x^j)$$
$$= \Sigma c_n^* x^n$$
$$= c^*(x)$$

(using the laws of S to collect together all terms involving x^n). Thus F preserves multiplication and is a homomorphism of rings. This proves the theorem.

COROLLARY 1. *For any commutative ring R and any element $x \in R$, the map from $R[X]$ to R, given by $a(X) \mapsto a(x)$, is a homomorphism of rings.*

Proof. This is just the special case of the theorem in which $R = S$ and f is the identity map.

COROLLARY 2. *(Remainder Theorem.) Let F be a field, let $t \in F$ and let $a(X) \in F[X]$. Then the remainder r on dividing $a(X)$ by $X - t$ is $a(t)$.*

Proof. By Theorem 10C we have

$$a(X) = (X - t)q(t) + r, \qquad (1)$$

where $\deg(r) < 1$, i.e., $r \in F$. By Corollary 1, the map which sends each polynomial $b(X)$ to $b(t) \in F$ is a homomorphism of rings and therefore preserves the relation (1) between polynomials. Hence

$$a(t) = (t - t)q(t) + r$$

in F, and so $r = a(t)$.

COROLLARY 3. *(Factor Theorem). Let F be a field, let $t \in F$ and let $a(X) \in F[X]$. Then $X - t$ divides $a(X)$ in $F[X]$ if and only if $a(t) = 0$ in F.*

Proof. $(X - t) \mid a(X) \Leftrightarrow r = 0$ in Corollary 2.

An element $t \in F$ such that $a(t) = 0$ is called a *root* of $a(X)$ in F. A given polynomial $a(X) \in F[X]$ may not have any roots in F. For example, $X^2 + 2$ has no root in **R** because $t^2 + 2 \geq 2$ for all $t \in \mathbf{R}$, so we cannot have $t^2 + 2 = 0$.

THEOREM 10K. *Let F be a field and let $a(X) \in F[X]$ have degree $n \geq 0$. Then $a(X)$ has at most n distinct roots in F.*

Proof. We use induction on n. If $n = 0$ then $a(X)$ is a non-zero element of F and has no roots. If $n > 0$ and $a(X)$ has no roots, the statement is true. If it has a root t, then by the factor theorem $a(X) = (X - t)b(X)$, where $b(X)$ has degree $n - 1$. If s is any other root of $a(X)$ in F with $s \neq t$, then $0 = a(s) = (s - t)b(s)$, whence $b(s) = 0$ since $s - t \neq 0$ and F is a field. But, by induction hypothesis, $b(X)$ has at most $n - 1$ distinct roots, so $a(X)$ has at most $n - 1$ distinct roots other than t, hence at most n altogether. The theorem now follows for all n.

COROLLARY 1. *If a polynomial $a(X) \in F[X]$ has degree less than m and has m distinct roots in F then it is the zero polynomial.*

Proof. If $a(X) \neq 0$ then it has degree $n \geq 0$ and so cannot have more than n distinct roots.

COROLLARY 2. *If $a(X), b(X) \in F[X]$ both have degree less than m and if $a(t) = b(t)$ for m distinct values of t in F, then $a(X) = b(X)$.*

Proof. Apply Corollary 1 to the polynomial $a(X) - b(X)$, whose degree is less than m.

COROLLARY 3. *Let F be an infinite field and let $a(X), b(X) \in F[X]$. If $a(t) = b(t)$ for all $t \in F$, then $a(X) = b(X)$. Thus distinct polynomials determine distinct polynomial functions in this case.*

Proof. Let n be the larger of the two degrees of $a(X)$ and $b(X)$. Then $a(t) = b(t)$ for $n + 1$ distinct values of t in F, so Corollary 2 implies that $a(X) = b(X)$.

COROLLARY 4. *Let F be any field and let $a(X) \in F[X]$ have degree $n > 0$. If $a(X)$ has n distinct roots $t_1, t_2, \ldots, t_n$ in F then $a(X) = u(X - t_1)(X - t_2) \ldots (X - t_n)$, where u is the leading coefficient of $a(X)$.*

Proof. Consider the polynomial
$$p(X) = a(X) - u \cdot (X - t_1)(X - t_2) \ldots (X - t_n).$$
Its degree is less than n because the coefficient of X^n is $u - u = 0$. But $b(X)$ has n distinct roots $t_1, t_2, \ldots, t_n$, so it must be the zero polynomial.

Example 10.5. Let $F = \mathbf{Z}_p$ where p is a prime number. Then F is a field and all the above results are true for the polynomial ring $\mathbf{Z}_p[X]$. Now the polynomial $X^p - X$ has p distinct roots in $\mathbf{Z}_p$, namely, all the elements of $\mathbf{Z}_p$, by Fermat's theorem. We can therefore apply Corollary 4 above, and deduce that

$$X^p - X = X(X-1)(X-2)\ldots(X-p+1) \text{ in } \mathbf{Z}_p[X].$$

For example, if $p = 3$, we have

$$X(X-1)(X-2) = X(X^2 - 3X + 2) = X(X^2 - 1) = X^3 - X.$$

We can, of course, cancel X on both sides to obtain

$$X^{p-1} - 1 = (X-1)(X-2)\ldots(X-p+1) \text{ in } \mathbf{Z}_p[X]$$

The symbols $1, 2, \ldots$ in this equation stand for elements of $\mathbf{Z}_p$. Now equality of polynomials means equality of coefficients of corresponding powers of X, so we may "equate coefficients". The terms of degree 0 give

$$-1 = (-1)(-2)\ldots(-p+1) \text{ in } \mathbf{Z}_p.$$

In other words, reverting to genuine integers, we have $-1 \equiv (-1)(-2)\ldots(-p+1) \pmod{p}$, which reduces to $(p-1)! \equiv (-1)^p \pmod{p}$. Since $(-1)^p = -1$ if p is odd, and $(-1)^2 \equiv -1 \pmod 2$, we can write this

$$(p-1)! \equiv -1 \pmod{p} \text{ for all prime numbers } p,$$

a result known as Wilson's theorem.

The analogy between integers and polynomials over a field F can be pursued much further than we have done so far, and we will indicate a few of the results which can be obtained in this way. Congruences can be defined in the polynomial ring $F[X]$ by writing $a(X) \equiv b(X) \pmod{p(X)}$ if $p(X) \mid \{a(X) - b(X)\}$. This is the same as saying that $a(X)$ and $b(X)$ are in the same additive coset of the ideal $I = p(X)F[X]$ consisting of all multiples of $p(X)$. Thus, just as for integers, the residue classes of polynomials modulo $p(X)$ form a commutative ring, namely the quotient ring $F[X]/I$.

THEOREM 10L. *Let F be a field, let $p(X) \in F[X]$ and let I be the principal ideal generated by $p(X)$. Then $F[X]/I$ is a field if and only if $p(X)$ is irreducible.*

Proof. Suppose that $p(X)$ is irreducible. We know that $R = F[X]/I$ is a commutative ring and we must show that its non-zero elements have inverses. In other words, if $a(X) \in F[X]$ and $p(X) \nmid a(X)$ we must show that there is a polynomial $b(X)$ such that $a(X)b(X) \equiv 1 \pmod{p(X)}$. But, as for integers, $(p(X), a(X)) = 1$ by Theorem 10G, so we can find

$r(X)$, $s(X)$ such that $p(X)r(X) + a(X)s(X) = 1$. The polynomial $s(X)$ then satisfies $a(X)s(X) \equiv 1 \pmod{p(X)}$. On the other hand, if $p(X)$ is not irreducible then either (i) it is 0, or (ii) it is a unit, or (iii) $p(X) = a(X)b(X)$ where $a(X)$, $b(X)$ are non-zero and of smaller degree than $p(X)$. In case (i) $F[X]/I \cong F[X]$ is not a field. In case (ii) $I = F[X]$, so $F[X]/I$ is the one-element ring and is not a field. In case (iii) the residue classes containing $a(X)$ and $b(X)$ are not zero, but their product is zero, so $F[X]/I$ is not even an integral domain.

Example 10.6. Let $F = \mathbf{R}$ and let $p(X) = X^2 + 1$. Then $p(X)$ is irreducible in $\mathbf{R}[X]$ (otherwise it would be the product of two factors of degree 1 and would have a root in $\mathbf{R}$). Hence the residue classes modulo $X^2 + 1$ form a field. To calculate in this field, we use polynomials with real coefficients but reduce modulo $X^2 + 1$ by putting $X^2 = -1$ whenever it occurs. Every polynomial is then congruent to a polynomial of the form $a + bX$ ($a, b \in \mathbf{R}$) and no two such polynomials are congruent. We add and multiply these as polynomials, putting $X^2 = -1$, and the reader will recognize that the new field is just the field of complex numbers; for if we replace the symbol X everywhere by i the calculations are exactly those in $\mathbf{C}$.

Example 10.7. Let $F = \mathbf{Z}_2$ and let $p(X) = X^2 + X + 1$. Then $p(X)$ is irreducible in $\mathbf{Z}_2[X]$ since otherwise it would be a product of factors of degree 1 and would have a root in $\mathbf{Z}_2$; this is not so since $p(0) = p(1) = 1$. Hence the residue classes of polynomials modulo $X^2 + X + 1$ form a field. Any polynomial is congruent to one of the form $a + bX$ ($a, b \in \mathbf{Z}_2$) and we calculate as for polynomials, but reducing the answer to the form $a + bX$ by writing $X^2 = -(X + 1) = X + 1$ whenever X^2 occurs. There are just four polynomials of the form $a + bX$, namely, 0, 1, X and $1 + X$. The addition and multiplication tables are

+	0	1	X	$1 + X$
0	0	1	X	$1 + X$
1	1	0	$1 + X$	X
X	X	$1 + X$	0	1
$1 + X$	$1 + X$	X	1	0

×	0	1	X	$1 + X$
0	0	0	0	0
1	0	1	X	$1 + X$
X	0	X	$1 + X$	1
$1 + X$	0	$1 + X$	1	X.

These are the same tables (with a change of notation) as those in Example 9.2, so we have now established that that example is indeed a field with four elements. This method can be used to construct a finite field of order q, where q is any prime-power.

THEOREM 10M. *(Chinese Remainder Theorem for Polynomials.) Let F be a field, let $p_1(X), p_2(X), \ldots, p_n(X)$ be pairwise coprime polynomials in $F[X]$, and let $c_1(X), c_2(X), \ldots, c_n(X)$ be arbitrary polynomials in $F[X]$. Then there is a polynomial $a(X)$ in $F[X]$ such that $a(X) \equiv c_i(X) \pmod{p_i(X)}$ for $i = 1, 2, \ldots, n$, and $a(X)$ is unique modulo $p(X) = p_1(X)p_2(X)\ldots p_n(X)$.*

Proof. The proof is exactly as for integers. First, put $q_1(X) = p_2(X)p_3(X)\ldots p_n(X)$. Then $p_1(X)$ is coprime to $q_1(X)$ and we can therefore write $1 = p_1(X)r_1(X) + q_1(X)s_1(X)$. The polynomial $a_1(X) = q_1(X)s_1(X)$ now satisfies

$$a_1(X) \equiv 1 \pmod{p_1(X)}$$

and $a_1(X) \equiv 0 \pmod{p_i(X)}$ for $i = 2, 3, \ldots, n$.

Similarly we can find $a_j(X)$ for $j = 1, 2, \ldots, n$ so that

$$a_j(X) \equiv 1 \pmod{p_j(X)}$$

and $a_j(X) \equiv 0 \pmod{p_i(X)}$ for $i \neq j$.

The polynomial $a(X) = \Sigma a_i(X)c_i(X)$ now satisfies all the congruences $a(X) \equiv c_i(X) \pmod{p_i(X)}$. If $b(X)$ is another solution then $a(X) \equiv b(X) \pmod{p_i(X)}$ for $i = 1, 2, \ldots, n$ and it follows that $a(X) \equiv b(X) \pmod{p(X)}$. Clearly also, any polynomial congruent to $a(X) \pmod{p(X)}$ is a solution.

Example 10.8. An interesting special case of the Chinese remainder theorem is the following. Let $t_1, t_2, \ldots, t_n$ be distinct elements of F. Then the monic polynomials $X - t_1, X - t_2, \ldots, X - t_n$ are distinct and irreducible, so they are pairwise coprime. If we put $p_i(X) = X - t_i$ and take $c_i(X)$ to be an element c_i of F, then the theorem says that there is a polynomial $a(X)$ satisfying $a(X) \equiv c_i (\mod(X - t_i))$ for $i = 1, 2, \ldots, n$. Now the remainder theorem (non-Chinese!) says that $a(X) \equiv c_i (\mod(X - t_i)) \Leftrightarrow a(t_i) = c_i$. Hence there is a polynomial $a(X)$ which takes the given values $c_1, c_2, \ldots, c_n$ at the given (distinct) points $t_1, t_2, \ldots, t_n$. Furthermore, the set of all such polynomials is a residue class modulo $p(X) = (X - t_1)(X - t_2)\ldots(X - t_n)$. Since $p(X)$ has degree n, there is a unique polynomial $a(X)$ of degree less than n satisfying $a(t_i) = c_i$ for $i = 1, 2, \ldots, n$ (namely, the remainder on dividing any one such polynomial by $p(X)$). We can actually use the method given in the proof of the theorem to write down this polynomial. First we must find a polynomial $a_i(X)$ satisfying $a_i(t_i) = 1$,

$a_i(t_j) = 0$ for $j \neq i$. The polynomial
$$b_i(X) = \prod_{j \neq i} (X - t_j)$$
is almost right. It has $b_i(t_j) = 0$ for $j \neq i$, and
$$b_i(t_i) = (t_1 - t_i)(t_2 - t_i) \ldots (t_{i-1} - t_i)(t_{i+1} - t_i) \ldots (t_n - t_i) = s_i,$$
say. Hence we may take
$$a_i(X) = s_i^{-1} \prod_{j \neq i} (X - t_j).$$
The required polynomial is now
$$a(X) = \sum_{i=1}^{n} c_i a_i(X) = \sum_{i=1}^{n} c_i s_i^{-1} \prod_{j \neq i} (X - t_j),$$
a formula first found by Lagrange. It is known as Lagrange's interpolation formula because when $F = \mathbf{R}$ it enables one to interpolate values of a function between the given values c_i at the points t_i. The polynomial $a(X)$ obviously has degree at most $n - 1$ and is therefore the unique one with this property.

The extension of $\mathbf{Z}$ to its field of fractions $\mathbf{Q}$ is a construction which works for any integral domain (Theorem 8J). In particular, if D is an integral domain, so is $D[X]$, by Theorem 10A, and therefore $D[X]$ has a field of fractions. The main interest lies in the case when $D = F$ is a field, and we denote the field of fractions of $F[X]$ by $F(X)$. Its members are fractions $a(X)/b(X)$, where $a(X)$ and $b(X)$ are polynomials and $b(X) \neq 0$. Such fractions are called *rational functions*, a very bad name since they are not functions at all, though they can be used to determine functions as in the case of polynomials. A better name would be "rational forms". Two fractions
$$\frac{a_1(X)}{b_1(X)} \quad \text{and} \quad \frac{a_2(X)}{b_2(X)}$$
are equal in $F(X)$ if and only if $a_1(X)b_2(X) = b_1(X)a_2(X)$ in $F[X]$. Because of this, common factors of numerator and denominator can be cancelled. If we express both numerator and denominator as products of units and monic irreducibles, and cancel where possible, we see that every element of $F(X)$ can be uniquely expressed in the form $u p_1(X)^{\alpha_1} p_2(X)^{\alpha_2} \ldots p_n(X)^{\alpha_n}$ where u is a unit, $p_1(X), p_2(X), \ldots, p_n(X)$ are distinct monic irreducible polynomials and $\alpha_i \in \mathbf{Z}, \alpha_i \neq 0$. Here, of course, we allow α_i to be negative to take care of the denominator. It is a good exercise to deduce the uniqueness of this factorisation in $F(X)$ from the known unique factorization theorem in $F[X]$.

The theory of partial fractions which was developed for **Q** in Chapter 9 works equally well in $F(X)$. (Indeed it is perhaps more familiar in this context because of its use in integrating rational functions of a real variable.) It would be pointless now to go through the same work again, so we simply state the main result and advise the reader to supply the proof (if possible without referring back to Chapter 9).

THEOREM 10N. *Let F be a field. Then every non-zero element of the rational function field $F(X)$ can be uniquely expressed in the form*

$$a(X) + \frac{b_1(X)}{p_1(X)^{\alpha_1}} + \frac{b_2(X)}{p_2(X)^{\alpha_2}} + \ldots + \frac{b_n(X)}{p_n(X)^{\alpha_n}},$$

where $a(X), b_i(X), p_i(X)$ are polynomials in $F[X]$, $p_1(X), p_2(X), \ldots, p_n(X)$ are monic irreducibles (not necessarily distinct), $\alpha_1, \alpha_2, \ldots, \alpha_n$ are positive integers, $p_1(X)^{\alpha_1}, p_2(X)^{\alpha_2}, \ldots, p_n(X)^{\alpha_n}$ are distinct, and $\deg(b_i(X)) < \deg(p_i(X))$ for $i = 1, 2, \ldots, n$.

Exercises

1. Use Euclid's algorithm to find the greatest common divisor in **R**$[X]$ of
$$X^5 - 2X^3 + X^2 - 3X + 1$$
and $\quad X^4 - 2X^3 - 3X^2 + 7X - 2$.

2. Find the greatest common divisor of $X^{12} + 1$ and $X^9 + 1$ in $\mathbf{Z}_3[X]$ and in $\mathbf{Z}_2[X]$.

3. Find the greatest common divisor of $X^{72} - 1$ and $X^{45} - 1$ in $\mathbf{Q}[X]$, and express it in the form $p(X)(X^{72} - 1) + q(X)(X^{45} - 1)$.

4. Prove that if a polynomial $f(X) \in F[X]$ of degree 3 has no roots in F then it is irreducible. Give an example to show that this is not true for polynomials of degree 4. (F is a field.)

5. Prove that if $a(X), b(X) \in F[X]$ and if F is a subfield of a field F', then the greatest common divisor of $a(X)$ and $b(X)$ in $F[X]$ is the same as their greatest common divisor in $F'[X]$. (N.B. This does not contradict the answers to Exercise 2 above since neither $\mathbf{Z}_2$ nor $\mathbf{Z}_3$ is a subfield of the other.)

6. Let F be a field. Let J be the set of all polynomials $a(X) \in F[X]$ such that $a(x) = 0$ for all $x \in F$. Prove that J is an ideal of $F[X]$ and find a generator for it (i) if $F = \mathbf{R}$, (ii) if $F = \mathbf{Z}_3$. (Note that by Theorem 10D, J is a principal ideal.)

7. Find a polynomial $a(X) \in F[X]$ of degree ≤ 3 such that $a(-1) = 0, a(0) = 3, a(1) = -2, a(2) = 1$ when (i) $F = \mathbf{Q}$ and (ii) $F = \mathbf{Z}_7$.

8. Show that the polynomials $X^3 + 2$ and $X^2 - X$ in $\mathbf{R}[X]$ are coprime, and find a real polynomial $p(X)$ such that
$$p(X) \equiv X - 1 \pmod{X^3 + 2}$$
$$p(X) \equiv 2X \pmod{X^2 - X}.$$
If the polynomials $X^3 + 2$ and $X^2 - X$ are considered as elements of $\mathbf{Z}_3[X]$, are they then coprime? Is there a polynomial $p(X) \in \mathbf{Z}_3[X]$ satisfying the given simultaneous congruences?

9. Let F be a field and let $p(X), q(X) \in F[X]$, where $\deg(q(X)) < \deg(p(X))$. Prove that there exist polynomials $p_0(X), p_1(X), \ldots, p_k(X)$, each of degree less than $\deg(q(X))$ such that $p(X) = p_0(X) + p_1(X)q(X) + p_2(X)(q(X))^2 + \ldots + p_k(X)(q(X))^k$. Show that the $p_i(X)$ are unique.

10. Let $r(X) = p(X)/[a(X) \cdot b(X)]$ in the rational function field $F(X)$, where F is a field. Let $a(X), b(X)$ be coprime polynomials and let $\deg(p(X)) < \deg(a(X)) + \deg(b(X))$. Prove that $r(X)$ can be written in the form
$$r(X) = \frac{c(X)}{a(X)} + \frac{d(X)}{b(X)}$$
where $c(X)$ and $d(X)$ are polynomials with $\deg(c(X)) < \deg(a(X))$ and $\deg(d(X)) < \deg(b(X))$.

11. Let $p(X) = p_0 + p_1 X + \ldots + p_n X^n$ where $p_i \in \mathbf{Z}$ and $p_n \neq 0$. Prove that if $p(t) = 0$, where $t \in \mathbf{Q}$, then $t = r/s$, where $r, s \in \mathbf{Z}$ and $r \mid p_0, s \mid p_n$.

12. Let p be a prime number. Prove that $(1 + X)^p = 1 + X^p$ in $\mathbf{Z}_p[X]$ and deduce that
$$(1 + X)^{p-1} = 1 - X + X^2 - \ldots + (-1)^{p-1} X^{p-1}.$$
Hence show that
$$p \mid \binom{p}{i} \text{ for } i = 1, 2, \ldots, p - 1$$
and that
$$\binom{p-1}{i} \equiv (-1)^i \pmod{p} \text{ for } i = 0, 1, 2, \ldots, p - 1.$$

13. Prove that the product of all elements of any Abelian multiplicative group is equal to the product of all the elements of order 2. Show that the multiplicative group of the field $\mathbf{Z}_p$ (p prime) has exactly one element of order 2 if $p > 2$ and

deduce Wilson's theorem:

$$(p - 1)! \equiv -1 \pmod{p}.$$

14. Prove that
$$\sum_{k=1}^{p-1} k^3 \equiv 0 \pmod{p}$$

if p is a prime number $\geqslant 5$. (Hint: the sum of the cubes of the roots of a polynomial can be expressed in terms of the coefficients.)

15. Prove that for any integer n and any prime number p,
$$\sum_{k=1}^{p-1} n^k \equiv 0 \text{ or } -1 \pmod{p}$$

16. Prove that if $2^n + 1$ is a prime number then n is a power of 2. (Hint: consider factorizations of the polynomial $X^n + 1$).

17. Prove that if p is an odd prime number then there are exactly $\frac{1}{2}(p - 1)$ elements in $\mathbf{Z}_p$ which are squares of non-zero elements in $\mathbf{Z}_p$. Show that these squares are the roots in $\mathbf{Z}_p$ of the polynomial $X^{\frac{1}{2}(p-1)} - 1$ and that the non-squares in $\mathbf{Z}_p$ are the roots of $X^{\frac{1}{2}(p-1)} + 1$. Deduce that the congruence

$$x^2 \equiv -1 \pmod{p},$$

where p is prime, has a solution if and only if $p \equiv 1 \pmod{4}$ or $p = 2$.

CHAPTER 11

Polynomials Over C, R, Q, and Z

The factorization of a polynomial $a(X) \in F[X]$ into irreducible factors may change if one passes to a larger field F' and asks for the irreducible factors in $F'[X]$. The relationship between the two factorizations is easily described.

THEOREM 11A. *Let F be a subfield of the field F'.*
(i) *If a polynomial $p(X) \in F[X]$ is irreducible in $F'[X]$ then it is irreducible in $F[X]$.*
(ii) *If a polynomial $a(X) \in F[X]$ has irreducible factorization $a(X) = up_1(X)p_2(X)\ldots p_n(X)$ in $F'[X]$ and if each $p_i(X)$ has coefficients in F then this is also the irreducible factorization of $a(X)$ in $F[X]$.*
(iii) *(General case). Let $a(X) \in F[X]$ have irreducible factorization $a(X) = vq_1(X)q_2(X)\ldots q_m(X)$ in $F[X]$. Then the irreducible factorization of $a(X)$ in $F'[X]$ can be obtained from it by further factorizing each $q_i(X)$ as a product of irreducible polynomials in $F'[X]$. Conversely, if $a(X)$ has irreducible factorization $a(X) = up_1(X)p_2(X)\ldots p_n(X)$ in $F'[X]$ and if each $p_i(X)$ is monic, then the irreducible factorization of $a(X)$ in $F[X]$ is of the form $uq_1(X)q_2(X)\ldots q_m(X)$, where, after suitable renumbering of $p_1, p_2, \ldots, p_n$, we have*

$$q_1(X) = p_1(X)p_2(X)\ldots p_i(X),$$
$$q_2(X) = p_{i+1}(X)p_{i+2}(X)\ldots p_j(X),$$
$$\cdots\cdots\cdots\cdots\cdots\cdots\cdots$$
$$q_m(X) = p_{k+1}(X)p_{k+2}(X)\ldots p_n(X).$$

Proof. (i) is obvious, and (ii) follows because if the p_i have coefficients in F they are irreducible in $F[X]$, and the unit u must also lie in F. (iii) Clearly, if we express each q_i as a product of irreducibles in $F'[X]$ then we obtain the required factorization in $F'[X]$ since the unit v of $F[X]$ remains a unit in $F'[X]$. For the converse, let $a(X) = up_1(X)\ldots p_n(X)$ in $F'[X]$, where the $p_i(X)$ are *monic* irreducibles. Then the unit u and the p_i are uniquely determined (apart from a permutation of the factors). Clearly $u \in F$ since it is the leading coefficient of $a(X)$. If we now compare this factorization with an irreducible factorization

$a(X) = vq_1(X) \ldots q_m(X)$ in $F[X]$ in which the q_j are monic, then we see immediately that $v = u$. Furthermore if we express each q_j as a product of monic irreducibles in $F'[X]$ (which can be done since q_j is monic) we must, by uniqueness of factorization in $F'[X]$ obtain precisely the factors $p_1(X) \ldots p_n(X)$ in some order. Thus the q_j are obtained by grouping together the p_i as indicated.

THEOREM 11B. *Let F be a subfield of the field F' and let $a(X), b(X) \in F[X]$. If $a(X) \mid b(X)$ in $F'[X]$, then $a(X) \mid b(X)$ in $F[X]$.*

Proof. If $a(X) = 0$, the result is trivial. If not, then by the Euclidean property of $F[X]$, we have $b(X) = a(X)q(X) + r(X)$, where $q(X)$, $r(X) \in F[X]$ and $\deg(r(X)) < \deg(a(X))$. All these polynomials are in $F'[X]$, and $a(X) \mid b(X)$ in $F'[X]$. It follows that $a(X) \mid r(X)$ in $F'[X]$. But $\deg(r(X)) < \deg(a(X))$, so $r(X)$ must be the zero polynomial by Theorem 10A. Hence $b(X) = a(X)q(X)$, where $q(X) \in F[X]$, as claimed.

We apply these theorems to compare factorizations in $\mathbf{R}[X]$ and $\mathbf{C}[X]$. Lacking a precise definition of $\mathbf{R}$ and $\mathbf{C}$ we clearly cannot prove much about them, so we assume (i) that $\mathbf{C}$ is a field with subfields $\mathbf{R} \supset \mathbf{Q}$ and (ii) that every element of $\mathbf{C}$ is uniquely of the form $a + bi$ with $a, b \in \mathbf{R}$, where i is a fixed element of $\mathbf{C}$ such that $i^2 = -1$. A proper description of these fields can be found in Birkhoff and MacLane, *A survey of Modern Algebra* (Macmillan, 1953). We shall also assume the following much deeper fact about $\mathbf{C}$ which is basic for factorization in $\mathbf{C}[X]$.

Fundamental Theorem of Algebra. *Every polynomial in $\mathbf{C}[X]$ of degree at least 1 has a root in $\mathbf{C}$.*

There are two main approaches to proving this theorem. The first uses the theory of analytic functions of a complex variable. This is a very beautiful theory which all students of mathematics should study at some time, but it would not be profitable to look at such a proof in isolation. A simplified version of such a proof, omitting the analytical details can be found in the admirable book by Birkhoff and MacLane cited above. The second approach is to use the fact that polynomials in $\mathbf{R}[X]$ of *odd* degree always have at least one root in $\mathbf{R}$. This fact is intuitively more obvious and can be proved using only a small amount of real analysis. The step from $\mathbf{R}$ to $\mathbf{C}$ can then be done purely algebraically (see, for example, Van der Waerden, *Modern Algebra*, Vol. I (Ungar, 1950)).

THEOREM 11C. *Every irreducible polynomial in $\mathbf{C}[X]$ has degree 1. Hence every non-zero polynomial $a(X) \in \mathbf{C}[X]$ has a factorization* *of the form*

$$a(X) = u(X - t_1)(X - t_2) \ldots (X - t_n)$$

where $u, t_1, t_2, \ldots, t_n \in \mathbf{C}, u \neq 0, n \geq 0$.

Proof. Let $p(X) \in \mathbf{C}[X]$ be irreducible. By definition, $\deg(p(X)) \geq 1$, so by the fundamental theorem, $p(X)$ has a root in $\mathbf{C}$, say $p(t) = 0$. By the factor theorem (Theorem 10J, Corollary 3), $(X - t) \mid p(X)$, and since $p(X)$ is irreducible it follows that $p(X) = u(X - t)$ for some unit u. Thus $\deg(p(X)) = 1$. The rest of the theorem now follows from the unique factorization theorem (Theorem 10H).

Unfortunately, knowing that a polynomial has a factorization with linear factors does not help one to find the factors. Indeed, there is in general no possibility of finding them in any precise sense. The best one can hope for is an algorithm which will compute closer and closer approximations to the roots of a given polynomial. Such algorithms exist but are tedious to perform by hand. However, the theoretical knowledge of the existence of a linear factorization is important in itself and we shall now work out its consequences for real polynomials.

If $c = a + bi$ is a complex number ($a, b \in \mathbf{R}$) then the *complex conjugate* of c is the complex number $\bar{c} = a - bi$. Conjugates satisfy the following laws, which are easily verified:

(i) $\overline{c_1 + c_2} = \bar{c}_1 + \bar{c}_2$, for all $c_1, c_2 \in \mathbf{C}$,

(ii) $\overline{c_1 c_2} = \bar{c}_1 \bar{c}_2$, for all $c_1, c_2 \in \mathbf{C}$,

(iii) $\bar{c} = c \Leftrightarrow c \in \mathbf{R}$, and

(iv) $\bar{\bar{c}} = c$ for all $c \in \mathbf{C}$.

The statements (i) and (ii), together with the fact that $\bar{1} = 1$, say that the map $c \mapsto \bar{c}$ from $\mathbf{C}$ to $\mathbf{C}$ is a homomorphism of rings. It is bijective (indeed, it is its own inverse) by (iv). Thus it is an isomorphism of $\mathbf{C}$ with itself. (Such an isomorphism is called an *automorphism* of $\mathbf{C}$.)

THEOREM 11D. *Let $a(X) \in \mathbf{R}[X]$ and let x be a root of $a(X)$ in $\mathbf{C}$. Then $\bar{x}$ is also a root of $a(X)$ in $\mathbf{C}$.*

Proof. For any polynomial $p(X) \in \mathbf{C}[X]$, say $p(X) = \Sigma p_i X^i$, we write $\bar{p}(X) = \Sigma \bar{p}_i X^i$. Since $c \mapsto \bar{c}$ is an automorphism, we have, for any complex number x, $\overline{(x^i)} = (\bar{x})^i$, $\overline{p_i x^i} = \bar{p}_i \bar{x}^i$, and hence $\overline{p(x)} = \bar{p}(\bar{x})$. In particular, if $a(X) \in \mathbf{R}[X]$ we have $\overline{a(x)} = a(\bar{x})$. Thus if $a(x) = 0$, it follows that $a(\bar{x}) = 0$ also.

COROLLARY 1. *Every irreducible polynomial in $\mathbf{R}[X]$ is either of degree 1 or is of the form $aX^2 + bX + c$ where $a, b, c \in \mathbf{R}$, and $b^2 < 4ac$. Conversely, all such polynomials are irreducible.*

Proof. Let $p(X) \in \mathbf{R}[X]$ be irreducible. Then $\deg(p(X)) \geq 1$, so $\exists\, x \in \mathbf{C}$ such that $p(x) = 0$ (by the Fundamental Theorem). If $x \in \mathbf{R}$, then, by the Factor Theorem, $(X - x) \mid p(X)$ in $\mathbf{R}[X]$, so $p(X) = u \cdot (X - x)$ where u is a unit in $\mathbf{R}[X]$ (because $p(X)$ is irreducible). Thus $p(X)$ has degree 1 in this case. If, on the other hand, $x \notin \mathbf{R}$, then $\bar{x} \neq x$ and $p(\bar{x}) = 0$. It follows from the Factor Theorem that $(X - x) \mid p(X)$ and $(X - \bar{x}) \mid p(X)$ in $\mathbf{C}[X]$. Hence $(X - x)(X - \bar{x}) \mid p(X)$ in $\mathbf{C}[X]$, because $X - x$ and $X - \bar{x}$ are coprime. Now

$$(X - x)(X - \bar{x}) = X^2 - (x + \bar{x}) + x\bar{x},$$

and if $x = s + ti$ then $x + \bar{x} = 2s$ and $x\bar{x} = s^2 + t^2$ are both real. So $p(X)$ is divisible in $\mathbf{C}[X]$ by a polynomial of degree 2 with real coefficients. According to Theorem 11B, $p(X)$ is still divisible by this polynomial in $\mathbf{R}[X]$ and therefore has degree 2 since it is irreducible in $\mathbf{R}[X]$. Thus $p(X) = a(X^2 - 2sX + (s^2 + t^2))$ where a, s and $t \in \mathbf{R}$ and $a \neq 0$. Writing $b = -2as$ and $c = a(s^2 + t^2)$ for the coefficients of X^1 and X^0, we have $b^2 - 4ac = 4a^2 s^2 - 4a^2(s^2 + t^2) = -4a^2 t^2 < 0$. Conversely, all polynomials of degree 1 are obviously irreducible. If $p(X) = aX^2 + bX + c$ with $b^2 < 4ac$, then $p(X)$ is also irreducible, since otherwise it would be a product of two factors of degree 1 and would therefore have a root $t \in \mathbf{R}$; but $at^2 + bt + c = 0$ implies $b^2 - 4ac = b^2 + 4a(at^2 + bt) = (2at + b)^2 \geq 0$, a contradiction.

COROLLARY 2. *Every polynomial of degree ≥ 1, with real coefficients, has a factorization as a product of polynomials of degree 1 or 2 with real coefficients.*

Proof. This follows from Corollary 1 and the Unique Factorisation Theorem.

Example 11.1. Consider the polynomial $a(X) = X^n - 1$ in $\mathbf{R}[X]$. We know that $a(X)$ has n distinct roots in $\mathbf{C}$, namely, $1, \zeta, \zeta^2, \ldots, \zeta^{n-1}$, where $\zeta = e^{2\pi i/n}$. These are all distinct, and therefore $X^n - 1 = (X - 1)(X - \zeta)(X - \zeta^2) \ldots (X - \zeta^{n-1})$ by Theorem 10K, Corollary 4. Now the root 1 is real, and so is the root $-1 = \zeta^{n/2}$ if n is even. The other roots occur, according to Theorem 11D, in conjugate pairs, each pair giving rise to an irreducible polynomial in $\mathbf{R}[X]$. If $n = 2m$ is even, the conjugate pairs are ζ^r and $\zeta^{-r} = \zeta^{n-r}$ for $r = 1, 2, \ldots m - 1$. The corresponding quadratic factors are $(X - \zeta^r)(X - \zeta^{-r}) = X^2 - (2\cos 2\pi r/n)X + 1$, so the irreducible factorization of $X^n - 1$ in $\mathbf{R}[X]$ is

$$X^n - 1 = (X - 1)(X + 1) \prod_{r=1}^{m-1} \left(X^2 - \left(2\cos \frac{2\pi r}{n} \right) X + 1 \right).$$

Similarly, if $n = 2m - 1$ is odd, we have

$$X^n - 1 = (X - 1) \prod_{r=1}^{m-1} \left(X^2 - \left(2\cos \frac{2\pi r}{n} \right) X + 1 \right).$$

Factorisation in $\mathbf{Q}[X]$ is more difficult. Of course, it sometimes happens that the irreducible factors in $\mathbf{R}[X]$ of a polynomial $a(X) \in \mathbf{Q}[X]$ all have coefficients in $\mathbf{Q}$, in which case they are also the irreducible factors in $\mathbf{Q}[X]$ (Theorem 11A). Failing this it is hard to tell whether a polynomial in $\mathbf{Q}[X]$ is irreducible or not. It is possible to test whether $a(X) \in \mathbf{Q}[X]$ has a root in $\mathbf{Q}$ or not and to find all such roots; but this only settles the irreducibility of polynomials up to degree 3, and $\mathbf{Q}[X]$ contains irreducible polynomials of arbitrarily high degree, as we shall see. The problem is best tackled by reducing it to questions of factorization in $\mathbf{Z}[X]$, and since $\mathbf{Z}$ is not a field we shall have to tread warily.

Irreducibility in $\mathbf{Z}[X]$ is defined exactly as before: $p(X) \neq 0$ is irreducible if it is not a unit in $\mathbf{Z}[X]$ and its only divisors are of the form u or $up(X)$ where u is a unit. Now the units in $\mathbf{Z}[X]$ are just the units of $\mathbf{Z}$, namely 1 and -1, so we now have a new phenomenon — any prime number in $\mathbf{Z}$ is an irreducible polynomial of degree 0 in $\mathbf{Z}[X]$. Another difficulty is that in $\mathbf{Z}[X]$, not every polynomial is the product of a unit and a monic polynomial. One must be careful, therefore, when comparing factorisations in $\mathbf{Z}[X]$ and $\mathbf{Q}[X]$.

An extremely useful device in dealing with factorisation in $\mathbf{Z}[X]$ is "reduction (mod p)". We fix a prime number p and consider the quotient map $\sigma : \mathbf{Z} \to \mathbf{Z}_p$. This is a homomorphism of rings, so it extends uniquely (by Theorem 10J) to a homomorphism from $\mathbf{Z}[X]$ to $\mathbf{Z}_p[X]$ sending X to X. This homomorphism is given by $\Sigma a_i X^i \mapsto \Sigma \sigma(a_i) X^i$ and is clearly surjective. Hence we have:

THEOREM 11E. *Let $p \in \mathbf{Z}$ be prime and, for any $a(X) \in \mathbf{Z}[X]$, let $a^*(X)$ be the polynomial whose coefficients are the residue classes (mod p) of the corresponding coefficients of $a(X)$. Then the map $a(X) \mapsto a^*(X)$ is a surjective ring homomorphism from $\mathbf{Z}[X]$ to $\mathbf{Z}_p[X]$.*

Given a polynomial $a(X) = \Sigma a_i X^i$ in $\mathbf{Z}[X]$ we define the *content* of $a(X)$ to be the greatest common divisor in $\mathbf{Z}$ of all its coefficients. The g.c.d. of $a_1, a_2, \ldots, a_n$ is defined inductively as

$$d = (\ldots ((a_1, a_2), a_3), \ldots, a_n).$$

It is easy to see that d is characterized by the properties:

(i) $d|a_i$ in $\mathbf{Z}$ for $i = 0, 1, 2, \ldots, n$;
(ii) if $c|a_i$ in $\mathbf{Z}$ for $i = 0, 1, 2, \ldots, n$, then $c|d$ in $\mathbf{Z}$;
(iii) $d \geqslant 0$.

We denote by $\gamma(a(X))$ the content of $a(X)$, and we observe that if $\gamma(a(X)) = d$, then $a(X) = d \cdot b(X)$, where $b(X) \in \mathbf{Z}[X]$ and $\gamma(b(X)) = 1$ (for if $\gamma(b(X)) = c$, then all coefficients of $a(X)$ are divisible by dc). We can now prove the crucial result for factorization in $\mathbf{Z}[X]$.

THEOREM 11F (*Gauss' Lemma*).   Let $a(X), b(X) \in \mathbf{Z}[X]$. Then $\gamma\{a(X)b(X)\} = \gamma(a(X))\,\gamma(b(X))$.

Proof. If one of the polynomials is 0, its content is 0, and the theorem is trivially true. So let $\gamma(a(X)) = r > 0$, $\gamma(b(X)) = s > 0$. Then $a(X) = r \cdot a_1(X)$ and $b(X) = s \cdot b_1(X)$, where $a_1(X)$ and $b_1(X)$ each have content 1. Thus $a(X)b(X) = rs \cdot a_1(X) b_1(X)$ has all its coefficients divisible by rs, and it will be enough to show that $\gamma\{a_1(X) b_1(X)\} = 1$. Suppose that this is false; then there is a prime number p which divides all coefficients of $a_1(X) b_1(X) = c(X)$, say. If we now reduce coefficients (mod p) for this prime, Theorem 11E gives

$$a_1^*(X)b_1^*(X) = c^*(X) = 0 \text{ in } \mathbf{Z}_p[X].$$

But $\mathbf{Z}_p[X]$ is an integral domain (Theorem 10A) since $\mathbf{Z}_p$ is a field, so we deduce that either $a_1^*(X) = 0$ or $b_1^*(X) = 0$ in $\mathbf{Z}_p[X]$. Returning to $\mathbf{Z}[X]_4$, this says that either $a_1(X)$ or $b_1(X)$ has all its coefficients divisible by p, which contradicts the fact that both these polynomials have content 1.

This argument is a very good example of the usefulness of the abstract method. The statement of the theorem involves only integers and polynomials and can be proved directly in $\mathbf{Z}[X]$, but only by some rather involved calculations with the coefficients of the product of two polynomials to analyse the circumstances under which they are all divisible by a prime p. However, reduction modulo p gets rid of the complications by using the argument that $\mathbf{Z}_p$ is a field and therefore $\mathbf{Z}_p[X]$ is an integral domain. The proof of this last fact, it will be recalled, only uses the *leading* coefficients of the polynomials, and this is where the simplification lies. An even better example of this process of simplification by reduction (mod p) is the proof of Eisenstein's criterion given below (Theorem 11J).

COROLLARY.   Let $a(X), b(X) \in \mathbf{Z}[X]$. If $a(X)|b(X)$ in $\mathbf{Q}[X]$ and if the content of $a(X)$ is 1, then $a(X)|b(X)$ in $\mathbf{Z}[X]$.

Proof. We are given that $b(X) = a(X)q(X)$, where $q(X)$ has rational coefficients. There is an integer $n \neq 0$ such that $q_1(X) = nq(X)$ has integer coefficients (for example, let n be the product of the denominators of the non-zero coefficients of $q(X)$). Hence $nb(X) = a(X) q_1(X)$ in $\mathbf{Z}[X]$. By Gauss' lemma, $n\gamma(b(X)) = \gamma(q_1(X))$, since $\gamma(a(X)) = 1$. Hence n divides the content of $q_1(X)$ and so

$$q(X) = \frac{1}{n} q_1(X)$$

has integer coefficients. Thus $a(X) | b(X)$ in $\mathbf{Z}[X]$.

Example 11.2. Let $b(X) \in \mathbf{Z}[X]$, say $b(X) = b_0 + b_1 X + \ldots + b_n X^n$. Suppose that $b(X)$ has a root $t = r/s$ in $\mathbf{Q}$. Then $(X - t) | b(X)$ in $\mathbf{Q}[X]$, so $(sX - r) | b(X)$ in $\mathbf{Q}[X]$. We can choose r and s to be coprime integers, so that $\gamma(sX - r) = 1$. It follows, by the above corollary, that $(sX - r) | b(X)$ in $\mathbf{Z}[X]$, that is,

$$b_0 + b_1 X + \ldots + b_n X^n = (-r + sX)(c_0 + c_1 X + \ldots + c_{n-1} X^{n-1}),$$

where $c_0, c_1, \ldots, c_{n-1} \in \mathbf{Z}$. Equating coefficients of X^0 and X^n, we find that $b_0 = -rc_0$ and $b_n = sc_{n-1}$. Thus $r | b_0$ and $s | b_n$ in $\mathbf{Z}$. Since b_0 and b_n have only a finite number of divisors we can try all rational numbers r/s, where $r | b_0$ and $s | b_n$, to see whether they are roots, and thus determine whether or not $b(X)$ has any roots in $\mathbf{Q}$. For example, if $b(X) = X^2 - 2$, the only possible roots in $\mathbf{Q}$ are $t = \pm 1, \pm 2$. Since none of these is a root, it follows that $X^2 - 2$ has no rational root. This proof that "$\sqrt{2}$ is irrational" looks simpler than the usual one, but this is only because the work has already been done in proving Gauss' lemma and its corollary.

Example 11.3. If $b(X) = 3X^3 - 2X^2 + X - 4$, the possible roots of $b(X)$ in $\mathbf{Q}$ are

$$t = \pm 1, \pm 2, \pm 4, \pm \frac{1}{3}, \pm \frac{2}{3}, \pm \frac{4}{3}.$$

None of these actually satisfies $b(t) = 0$, so $b(X)$ has no roots in $\mathbf{Q}$. It is therefore irreducible in $\mathbf{Q}[X]$ since otherwise it would be the product of two polynomials of degrees 1 and 2 and so would have a root in $\mathbf{Q}$. It is also irreducible in $\mathbf{Z}[X]$ since it has no divisors $sX - r$ in $\mathbf{Z}[X]$ and has no divisors of degree 0 except ± 1 since its content is 1.

THEOREM 11G. *Let $a(X) \in \mathbf{Z}[X]$ have content 1. Then $a(X)$ is irreducible in $\mathbf{Z}[X]$ if and only if it is irreducible in $\mathbf{Q}[X]$.*

Proof. First let $a(X)$ be irreducible in $\mathbf{Q}[X]$. Then deg $(a(X)) \geq 1$, so $a(X)$ is not a unit in $\mathbf{Z}[X]$. If $a(X) = b(X) c(X)$ is any factorization in $\mathbf{Z}[X]$, then it is also a factorisation in $\mathbf{Q}[X]$, so one of the factors, say $b(X)$, is a unit in $\mathbf{Q}[X]$. Since $b(X)$ has integer coefficients it is an integer and must be ± 1 since $a(X)$ has content 1. Hence $a(X)$ is irreducible in $\mathbf{Z}[X]$.

Conversely, suppose that $a(X)$ is irreducible in $\mathbf{Z}[X]$. It is not zero and is not ± 1, so it must have degree $\geqslant 1$ since its content is 1. Suppose that $a(X) = b(X)c(X)$ in $\mathbf{Q}[X]$. There is a rational number $r \neq 0$ such that $b_1(X) = rb(X)$ has integer coefficients and content 1 (find an integer multiple of $b(X)$ which has integer coefficients and then divide by its content). Clearly $b_1(X) \mid a(X)$ in $\mathbf{Q}[X]$ and therefore, by the corollary to Theorem 11F, $b_1(X) \mid a(X)$ in $\mathbf{Z}[X]$. Since $a(X)$ is irreducible in $\mathbf{Z}[X]$, it follows that $b_1(X) = \pm 1$ or $b_1(X) = \pm a(X)$. Thus $b(X)$ is $\pm r^{-1}$ or $\pm r^{-1} a(X)$ and so $a(X)$ is irreducible in $\mathbf{Q}[X]$.

COROLLARY. *Let $a(X) \in \mathbf{Z}[X]$ have content n. Then $a(X)$ is irreducible in $\mathbf{Q}[X]$ if and only if $n^{-1} a(X)$ is irreducible in $\mathbf{Z}[X]$.*

THEOREM 11H. *Every non-zero polynomial $a(X) \in \mathbf{Z}[X]$ has a factorization*
$$a(X) = n \cdot p_1(X) p_2(X) \ldots p_r(X),$$
where $n \in \mathbf{Z}$ and the $p_i(X)$ are irreducible polynomials in $\mathbf{Z}[X]$ of content 1 with positive leading coefficients. The factorization is unique apart from the order of the factors.

Proof. In $\mathbf{Q}[X]$, Theorem 10H guarantees a factorization $a(X) = u \cdot q_1(X) q_2(X) \ldots q_r(X)$, where u is a non-zero rational number and each $q_i(X)$ is irreducible in $\mathbf{Q}[X]$. For suitable non-zero rational numbers r_i, the polynomials $p_i(X) = r_i q_i(X)$ have integer coefficients and content 1. Changing r_i to $-r_i$ if necessary, we may assume that the leading coefficient of $p_i(X)$ is positive. We now have $a(X) = v \cdot p_1(X) p_2(X) \ldots p_r(X)$ in $\mathbf{Q}[X]$ and we wish to show that the rational number v is an integer. This follows from the corollary to Theorem 11F, because $p_1(X) p_2(X) \ldots p_r(X)$ has content 1.

The uniqueness of the factorization is easy to prove. First, n is unique since, by Gauss' lemma, it is equal to $\pm \gamma(a(X))$ and its sign is that of the leading coefficient of $a(X)$. Second, each $p_i(X)$ is irreducible in $\mathbf{Q}[X]$ by Theorem 11G. Hence, if $a(X) = n p_1'(X) p_2'(X) \ldots p_s'(X)$ is another factorization of the given type then, by the unique factorization theorem for $\mathbf{Q}[X]$, $r = s$ and (after renumbering if necessary) $p_i(X) \mid p_i'(X)$ and $p_i'(X) \mid p_i(X)$ in $\mathbf{Q}[X]$. Since $p_i(X)$ and $p_i'(X)$ both have content 1, another application of the corollary to Theorem 11F shows that each divides the other in $\mathbf{Z}[X]$ and hence $p_i(X) = \pm p_i'(X)$. But both polynomials have positive leading coefficient, so they are equal, and the factorization is unique.

Example 11.4. The polynomial $X^6 + 1 \in \mathbf{Z}[X]$ has three pairs of conjugate roots in $\mathbf{C}$, namely, $\zeta^{\pm 1}, \zeta^{\pm 3}, \zeta^{\pm 5}$, where $\zeta = e^{2\pi i/12}$. The irreducible factors of $X^6 + 1$ in $\mathbf{R}[X]$ are therefore

and
$$(X - \zeta)(X - \zeta^{-1}) = X^2 - 2\cos\frac{\pi}{6} + 1,$$
$$(X - \zeta^3)(X - \zeta^{-3}) = X^2 + 1$$
$$(X - \zeta^5)(X - \zeta^{-5}) = X^2 - 2\cos\frac{5\pi}{6} + 1.$$

By Theorem 11A, the irreducible factors of $X^6 + 1$ in $\mathbf{Q}[X]$ are obtained by combining the above factors appropriately. Now $X^6 + 1 = (X^2 + 1)(X^4 - X^2 + 1)$ in $\mathbf{Q}[X]$, and

$$X^4 - X^2 + 1 = (X^2 - 2\cos\frac{\pi}{6} + 1)(X^2 - 2\cos\frac{5\pi}{6} + 1)$$

in $\mathbf{R}[X]$ so the only question is whether $\cos \pi/6$ is rational or not. But

$$\cos\frac{\pi}{6} = \frac{\sqrt{3}}{2}$$

is a root of the polynomial $4X^2 - 3$ which has no rational roots by the argument used in Examples 11.2 and 11.3. Hence the irreducible factors of $X^6 + 1$ in $\mathbf{Q}[X]$ are $X^2 + 1$ and $X^4 - X^2 + 1$. Both these are in $\mathbf{Z}[X]$ and have content 1, so they are also the irreducible factors of $X^6 + 1$ in $\mathbf{Z}[X]$.

We now prove one of the few known criteria for irreducibility of a polynomial in $\mathbf{Z}[X]$. It is a sufficient, but not necessary condition for irreducibility, so it is not powerful enough to decide in general whether a given polynomial is irreducible or not. However, it is very useful in some special cases and it shows, in particular, that there are irreducible polynomials in $\mathbf{Z}[X]$ of arbitrarily high degree.

THEOREM 11J. *(Eisenstein's Criterion.)* Let

$$a(X) = a_0 + a_1 X + \ldots + a_n X^n$$

be a polynomial of degree $n \geq 1$ with integer coefficients and content 1. Suppose that there exists a prime number p such that

(i) $p \mid a_i$ for $i = 0, 1, \ldots, n - 1$,
(ii) $p^2 \nmid a_0$.

Then $a(X)$ is irreducible in $\mathbf{Z}[X]$ and hence in $\mathbf{Q}[X]$.

Proof. We suppose that $a(X)$ is not irreducible in $\mathbf{Z}[X]$ and look for a contradiction. Since the content of $a(X)$ is 1 it has no irreducible factors of degree zero, so it must have a factorization

$$a(X) = b(X)c(X)$$

where $b(X), c(X) \in \mathbf{Z}[X]$ have degrees r, s, respectively, satisfying

$r \geq 1, s \geq 1, r+s = n$. We now reduce coefficients modulo p (the given prime) to obtain the equation $a^*(X) = b^*(X)c^*(X)$ in $\mathbf{Z}_p[X]$ (see Theorem 11E). By condition (i) we have $a^*(X) = uX^n$, where u is the residue class modulo p of a_n, and clearly $u \neq 0$ in $\mathbf{Z}_p$ since otherwise p would divide *all* the coefficients of $a(X)$.

Now $\mathbf{Z}_p$ is a field (Theorem 9A, Corollary) and therefore the unique factorization theorem is true in $\mathbf{Z}_p[X]$ (Theorem 10H). The polynomial X is irreducible and it follows that the only divisors of uX^n are of the form tX^m, where t is a non-zero element of $\mathbf{Z}_p$ and $m \leq n$. Since $b^*(X)$ and $c^*(X)$ have degrees at most r and s, respectively, the factorization $uX^n = b^*(X)c^*(X)$ must be of the form $uX^n = (vX^r)(wX^s)$. Returning to $\mathbf{Z}[X]$ we now see that the "constant terms" of $b(X)$ and $c(X)$, i.e., the terms b_0, c_0 of degree 0, are both divisible by p, because in $b^*(X)$ and $c^*(X)$ they become 0 (note that $r \geq 1, s \geq 1$). Hence $a_0 = b_0 c_0$ is divisible by p^2, contrary to condition (ii). This contradiction proves the theorem. (Irreducibility in $\mathbf{Q}[X]$ follows by Theorem 11G.)

Example 11.5. For any prime number p and any $n \geq 1$ the polynomial $X^n - p$ has content 1 and satisfies the Eisenstein criterion. Hence it is irreducible in $\mathbf{Z}[X]$ and in $\mathbf{Q}[X]$. This shows in particular that the positive nth root of p in $\mathbf{R}$ is not rational, a fact which can also be deduced easily from the unique factorization theorem for $\mathbf{Z}$.

Example 11.6. Let p be a prime number and consider the polynomial

$$a(X) = \frac{1}{X}\{(X+1)^p - 1\}$$
$$= X^{p-1} + pX^{p-2} + \ldots + \binom{p}{i}X^{p-i-1} + \ldots + p$$

in $\mathbf{Z}[X]$. It has content 1, since it is monic, and all the coefficients

$$\binom{p}{i} \text{ for } i = 1, 2, \ldots, p-1$$

are divisible by p. This may be seen from the formula

$$\binom{p}{i} = \frac{p!}{i!(p-i)!}$$

in which the numerator is divisible by p while the denominator is not. [Alternatively (see Exercise 12 of Chapter 10) one may argue that in $\mathbf{Z}_p[X]$ the polynomial $(X+1)^p - X^p - 1$ has degree $p-1$, and it has p distinct roots in $\mathbf{Z}_p$ because, for any $x \in \mathbf{Z}_p$,

$$(x+1)^p = x+1 = x^p + 1.$$

Hence this polynomial is the zero polynomial. It follows that in $\mathbf{Z}[X]$, all coefficients of $(X+1)^p - X^p - 1$ are divisible by p.] Returning to

$a(X)$ above, we see that its constant term p is not divisible by p^2 and therefore $a(X)$ is irreducible in $\mathbf{Z}[X]$ and in $\mathbf{Q}[X]$ by Eisenstein's criterion.

We end by comparing the factorisations in $\mathbf{C}[X]$, $\mathbf{R}[X]$ and $\mathbf{Q}[X]$ of the polynomials $X^n - 1$. In $\mathbf{C}[X]$ we have

$$X^n - 1 = \prod_{r=1}^{n} (X - \zeta^r),$$

where $\zeta = e^{2\pi i/n}$. In $\mathbf{R}[X]$ we obtain the irreducible factors by pairing off the conjugate roots ζ^r and $\zeta^{n-r} = \zeta^{-r}$ (see Example 11.1). To obtain a factor of $X^n - 1$ in $\mathbf{Q}[X]$ we must find a set of nth roots of 1 such that the product of the corresponding factors $X - \zeta^r$ has rational coefficients. An examination of low values of n suggests the solution of this problem. For example

$$X^6 - 1 = (X^3 - 1)(X^3 + 1)$$
$$= (X - 1)(X + 1)(X^2 + X + 1)(X^2 - X + 1).$$

The roots of these four factors, expressed in terms of $\zeta = e^{\pi i/3}$ are $\{\zeta^6\}, \{\zeta^3\}, \{\zeta^2, \zeta^4\}$ and $\{\zeta^1, \zeta^5\}$. These are respectively the primitive 1st roots, the primitive square roots, the primitive cube roots and the primitive 6th roots of 1. The reader should check that a similar pattern exists for other small values of n. This pattern suggests the following definition. For any positive integer m we define the mth *cyclotomic polynomial* to be the monic polynomial $\Phi_m(X)$ whose roots in $\mathbf{C}$ are the primitive mth roots of 1. Thus

$$\Phi_m(X) = \prod_{\rho \in S_m} (X - \rho)$$

where S_m is the set of all primitive mth roots of 1. The members of S_m are of the form ζ^r, where $\zeta = e^{2\pi i/m}$ and r runs through all the integers in the range $1 \leq r \leq m$ which are coprime to m (see Example 7.7). Hence S_m has $\phi(m)$ members and therefore the degree of $\Phi_m(X)$ is $\phi(m)$.

Now, for any $n \geq 1$, the set of *all* nth roots of 1 is the disjoint union of the sets S_m for all divisors m of n. It follows that

$$X^n - 1 = \prod_{m|n} \Phi_m(X)$$

in $\mathbf{C}[X]$. (We have simply grouped the linear factors $X - \rho$ of $X^n - 1$ according to the orders of the roots ρ.)

THEOREM 11K. *The cyclotomic polynomials $\Phi_m(X)$ have integer coefficients.*

Proof. We use induction on m. The polynomial $\Phi_1(X) = X - 1$

certainly has integer coefficients, so we assume that $\Phi_r(X)$ has integer coefficients for all $r < m$ and consider $\Phi_m(X)$. We have $X^m - 1 = \Phi_m(X)\Psi(X)$, where $\Psi(X)$ is the product of all $\Phi_r(X)$ with $r \mid m$ and $r \neq m$. By induction hypothesis, all these $\Phi_r(X)$ have integer coefficients and therefore so does $\Psi(X)$. Since $\Psi(X) \mid X^m - 1$ in $\mathbf{C}[X]$, Theorem 11B shows that $\Psi(X) \mid X^m - 1$ in $\mathbf{Q}[X]$, and therefore the quotient $\Phi_m(X)$ has rational coefficients. Furthermore, $\Psi(X)$ and $X^m - 1$ both lie in $\mathbf{Z}[X]$ and have content 1 since they are monic; therefore $\Psi(X) \mid X^m - 1$ in $\mathbf{Z}[X]$, by Theorem 11F, Corollary. Hence the quotient $\Phi_m(X)$ has integer coefficients, and the induction is complete.

We now have a standard factorization of $X^n - 1$ in $\mathbf{Z}[X]$, namely

$$X^n - 1 = \prod_{d \mid n} \Phi_d(X).$$

Notice that the formula for the degree of a product gives the familiar equation

$$n = \prod_{d \mid n} \phi(d).$$

It is a fact, too difficult to prove here, that all the cyclotomic polynomials are irreducible in $\mathbf{Z}[X]$ and in $\mathbf{Q}[X]$, so that we actually have the complete factorization of $X^n - 1$ in all cases. The interested reader will find a proof of this fact in Van der Waerden, *Modern Algebra*, Vol. I, (Ungar, 1950). We can, however, prove the following special case.

Example 11.7. If p is a prime number then the primitive pth roots of 1 are all the pth roots of 1 except 1 itself. Hence

$$\Phi_p(X) = \frac{X^p - 1}{X - 1} = X^{p-1} + X^{p-2} + \ldots + X + 1.$$

To see that this is irreducible in $\mathbf{Z}[X]$ we take a new symbol Y and consider the polynomial

$$a(Y) = \Phi_p(Y + 1) = Y^{-1}\{(Y + 1)^p - 1\} \in \mathbf{Z}[Y].$$

We proved in Example 11.6 that $a(Y)$ is irreducible in $\mathbf{Z}[Y]$, using the Eisenstein criterion. We can deduce that $\Phi_p(X)$ is irreducible in $\mathbf{Z}[X]$ by considering the map from $\mathbf{Z}[X]$ to $\mathbf{Z}[Y]$ which sends any polynomial $c(X)$ to $c(Y + 1)$. This map is a homomorphism of rings by Theorem 10J, and it sends each integer to itself. Hence any factorization $\Phi_p(X) = f(X)g(X)$ in $\mathbf{Z}[X]$ yields a factorisation $a(Y) = f(Y + 1)g(Y + 1)$ in $\mathbf{Z}[Y]$. One of the latter factors must be ± 1, so the same is true of the factors $f(X), g(X)$ in $\mathbf{Z}[X]$. This shows that $\Phi_p(X)$ is irreducible when p is prime.

Exercises.

1. Find the irreducible factors of $X^n + 1$ in $\mathbf{R}[X]$.
2. Find the irreducible factors of $X^{2n} + X^n + 1$ in $\mathbf{R}[X]$.
3. Express $X^{12} - 1$ as a product of irreducible polynomials in (i) $\mathbf{R}[X]$, (ii) $\mathbf{Q}[X]$, (iii) $\mathbf{Z}_2[X]$ and (iv) $\mathbf{Z}_7[X]$. Justify your answers.
4. Prove that the greatest common divisor in $\mathbf{Q}[X]$ of $X^m - 1$ and $X^n - 1$ is $X^d - 1$ where d is the greatest common divisor in $\mathbf{Z}$ of m and n.
5. Find the coefficients of $\Phi_9(X)$, $\Phi_{10}(X)$ and $\Phi_{36}(X)$.
6. Prove that $\Phi_n(0) = 1$ for $n \geq 2$.
7. Assuming the irreducibility of the cyclotomic polynomials, find the irreducible factors in $\mathbf{Z}[X]$ of $X^n + 1$ and $X^{2n} + 1$.

Index

Abelian group, 40
abstract algebra, 6
addition modulo n, 82
additive group, 41
    powers, 46
algebra of sets, 3
antisymmetric, 24
associative law, 5
automorphism, 145
axiomatic method, 6

bijection, 18
binary operation, 22
binomial theorem, 97
Boolean algebra, 5
Boole, George, 5

C, 12
C*, 41
cancellation law, 31
cardinal, 21
Chinese remainder theorem, 87
Chinese remainder theorem for
    polynomials, 138
circle group, 42
commutative group, 40
    law, 5
    ring, 96
complement, 13
complex conjugate, 145
    number, 3
composite function, 17
congruence, 37
content of a polynomial, 147
coprime integers, 61
    polynomials, 129
correspondence, 24
coset, 53
countable set, 21
cyclic group, 48

distributive law, 5
divides, 37
domain, 15

Eisenstein's criterion, 151
element, 11
empty set, 12
equality of functions, 16
    of sets, 12
equivalence class, 26
    relation, 25
Euclidean domain, 128
    group, 42
    property, 36
    property of polynomials, 126
Euclid's algorithm, 64, 129
Euler's function, 89
    theorem, 113
exponential function, 20

factor theorem, 134
Fermat's theorem, 114
fibres of a function, 26
field, 105
field of fractions, 109
    of rational numbers, 116
finite field, 113
    set, 21
first isomorphism theorem for
    groups, 77
    for rings, 103
fraction, 107
function, 15
    exponential, 20
    inverse of, 18
    logarithmic, 20
    of several variables, 21
    values of, 16
functions, composite of, 17
    equality of, 16

fundamental theorem of algebra, 144
of arithmetic, 62

Gaussian integers, 71
Gauss' lemma, 148
general linear group, 43
generator of a group, 48
of an ideal, 111
greatest common divisor, 59, 128
group, 40
    Abelian, 40
    additive, 41
    commutative, 40
    cyclic, 48
    Euclidean, 42
    finite, 41
    general linear, 43
    infinite, 41
    multiplicative, 41
    of units, 99
    order of, 41
    orthogonal, 42
    symmetric, 42
groups, isomorphic, 50

homomorphism of groups, 76
of rings, 101

ideal of a ring, 102
identity element, 41
    function, 18
implication, 15
inclusion map, 18
indexing set, 13
induction hypothesis, 34
induction, principle of, 34
infinite set, 21
injection, 18
integral domain, 104
intersection, 13
inverse, 40
    functions, 18

irreducible polynomials, 131
isomorphism class, 51
    of groups, 50
    of rings, 101
    theorems, 77
    type, 51

kernel of ring homomorphism, 103
of group homomorphism, 77
Klein 4-group, 71

Lagrange's interpolation formula, 139
Lagrange's theorem, 54
law, associative, 5
    commutative, 5
    distributive, 5
leading coefficient, 125
    term, 125
left coset, 53
linear congruence, 84
    order, 25
logarithmic function, 20

map, 17
mapping, 17
matrix algebra, 2
member, 11
modulo, 37
multiplicative group, 41

N, 12
natural numbers, 12
neutral element, 40
normal subgroup, 75

one-one correspondence, 18
operation, 1, 22
    binary, 1, 22
    unary, 1, 22
operations on quotient sets, 72
ordered pair, 21
order of a group, 41
    of an element, 49
    relation, 25
orthogonal group, 42
    transformation, 42

partial fractions, 117, 140
partition, 25
permutation, 42
polynomial, 2, 123
    function, 133
    irreducible, 131
    monic, 127
polynomials, coprime, 129

powers in a group, 45
prime number, 35, 62
primitive root, 90
principal ideal, 111
    ideal theorem, 127
principle of induction, 34
product of groups, 69
    of sets, 21

Q, 12
Q^*, 41
Q^{pos}, 41
quantifiers, 14
quotient group, 72
    map, 27, 103
    ring, 102
    set, 27

R, 12
R^*, 41
R^{pos}, 41
range, 15
rational function, 139
reduction (mod p), 147
reflexive, 24
relation, 24
    antisymmetric, 24
    equivalence, 25
    order, 25
    reflexive, 24
    symmetric, 24
    transitive, 24
relatively prime, 61
remainder, 37
    theorem, 134
residue, 37
    class, 37
right coset, 53
ring, 95
rings, isomorphic, 101
roots of a polynomial, 134

set, 11
    countable 21

empty, 12
finite, 21
indexing, 13
infinite, 21
sets, equality of, 12
    intersection of, 13
    product of, 21
    similar, 21
    union of, 13
similar sets, 21
simultaneous congruences, 85
$\mathscr{S}_n$, 42
standard algebra, 1
subgroup, 48
    normal, 75
subring, 99
subset, 12
surjection, 18
symmetric relation, 24
    group, 42
symmetry, 42

torsion-free, 80
total order, 25
transitive, 24

unary operation, 22
union, 13
unique factorization theorem
    for polynomials, 132
    for Z, 62
unit, 59, 98
universal property, 27

values of a function, 16
vector algebra, 3

well-ordering principle, 31
Wilson's theorem, 136

Z, 12
Z_n, 38, 111
zero-divisor, 103
zero element, 41